Wonderful World of Cupcakes

小杯子裏
嘆世界

戚黛黛・蒙順意 編著
Priscilla Chi・Monica Mong

萬里機構・飲食天地出版社

自序

從小到大我都好喜歡烘焙，不論是曲奇餅還是蛋糕，我都樂於嘗試。像很多小朋友一樣，我小的時候就喜歡扮演大人的角色，照顧別人的起居飲食，最直接的方法就是做一些甜點給大家吃。看到家人朋友品嚐甜品的幸福模樣，我總是非常開心！長大以後，幸福對我來講就是可以飛到世界各地品嚐美食，了解每個地方獨有的飲食文化和烹飪方法。為了製作出更健康美味的甜品，我跟隨不少老師學習烘焙，並完成了HKUSPACE為期8個月的營養學課程。

旅遊和烘焙是我人生中的兩大樂趣，我非常樂意將這份快樂與大家分享，於是就有了這本《小杯子裏嘆世界》！在這本書中，我選擇了十個我去過並留有深刻印象的地方，運用當地的美食或獨特的材料，創造出我心目中代表這個地區的cupcake。每個地方和每個cupcake我都寫了一段文字，介紹它們對我的意義或這個蛋糕的特別之處。我還邀請了與我一樣喜愛做蛋糕、對蛋糕裝飾頗有經驗與心得的好友Monica，與我一起研究cupcake的味道和裝飾。

我希望通過這本書，描繪出一張世界美食地圖，讓大家足不出戶，也可以跟隨不同口味的cupcake感受不同地方的飲食文化！我更希望大家能帶着輕鬆的心態跟着我和Monica一起動手做cupcake！你會發現，即使用基礎的材料和簡單的方法，只要花多一點心思去做裝飾，也可以做出讓人賞心悅目的cupcake!

戚黛黛

自序

多年前一個偶然機會下與朋友一起去學做蛋糕，便迷戀上了這個甜品陷阱不能自拔，自始對烘焙蛋糕產生了濃厚興趣。我慢慢地由基本蛋糕製作開始，一路跟隨不同資深甜品師學習，並不斷與朋友分享心得和成果。

最開心的是在朋友時生日送上獨一無二的自家製蛋糕，朋友收到蛋糕時的喜悅給了我很大鼓舞，讓曾當過時裝設計師的我重拾創作的快樂！為了提升自己，做出更具設計感的蛋糕，我完成了美國著名烘焙學校惠而通(Wilton School)的蛋糕裝飾技巧課程以及英國PME糖藝專業文憑課程。在與不同外國糖藝師學習的過程中，我深深地體會到蛋糕裝飾的魅力！

糖藝師就好像魔術師，將原本平淡無奇的蛋糕變成藝術品。而好的蛋糕不僅要求美味，也講究造型、構思、色彩和對主題的呼應，蛋糕裝飾的環節更可充分發揮個人創意，給創作者帶來莫大的滿足。

已經取得惠而通糖藝教師資格的我，希望將所學到的蛋糕裝飾技巧跟讀者們一起分享。只要有興趣，你也可以動手製作一個專業級的精緻蛋糕！

蒙順意

推薦序

由參選香港小姐到現在，我與黛黛相識已超過十年。

過往經常到她家中做客的我，有幸一嚐她非凡的廚藝，由她主理的甜品尤其不能錯過。

黛黛對創製甜品的熱誠多年來有增無減：她會特意到世界各地找尋關於甜品的食譜或靈感；在友人的聚會上，她會親自炮製各式各樣主題獨特的甜品，比如生日蛋糕、cupcake、曲奇餅等等，為在場的朋友送上一份甜絲絲的祝福。她亦常常在互聯網上分享新菜式，除了賣相精緻外，味道亦令人一試難忘。

得知她要出書，我當然替她高興。書裏除了分享做甜品的心得、富創意的食譜外，還有她在世界各地尋食覓味的經歷！最有趣的是，當你跟着這本書做甜品，感覺就好像跟着黛黛一起周遊列國尋找美食一樣！

Don't forget to bring your passport!

楊洛婷
2003 年香港小姐亞軍及著名活動司儀

認識了黛黛超過十年，知道她一路以來都非常熱愛烹飪，而且充滿藝術感。去過她家的朋友，都會知道她家裏每一個角落都充滿色彩，還陳列了來自世界各地的有趣裝飾品。食過她做的蛋糕，無論是味道或者賣相，都很有心思和創意。對

食物要求高又喜歡甜品和烹飪的我，看到這本書真的很興奮！我終於可以得到黛黛的秘笈！多謝你令我的烹飪世界裏增加了很多色彩！

李施嬅

2003年香港小姐最上鏡小姐及著名演員

當你感到情緒低落、心情鬱悶的時候，你會選擇用什麼方法去令自己開心？有些人會瘋狂購物，有些會睇書、看海、聽音樂、睡覺、洗澡、唱歌、漫無目的四處逛……總之各適其適。

但我相信有一種方法應該是最多人用，就是食，尤其是甜品或者蛋糕。當你見到一件精美的蛋糕放在眼前，還未品嚐，那種賞心悅目的感覺就已經將你的開心指數上升了一半；跟着將一小口放入口中，從舌尖一直刺激你的感官、神經，頓時就會使你變得開懷。為何簡單的材料就有這神奇的效果？有沒有想過要感謝為你製造出這件蛋糕的天使？慶幸曾

經有這樣的天使出現過在我身邊，她就是戚黛黛。

希望大家透過她這本書，找到開心、快樂的泉源！

蔡國威

六合彩王子

絲帶、蝴蝶結、手袋、cupcake 都是代
表黛黛。

以我非官方估計，黛黛應該是受了我
的 inspiration（又或是刺激）——連行外
人都做了餅店，真正的行家怎麼能不
出手？所以她終於將一本極具創意，
很有風土特色的 cupcake recipe 獻給大
家。當然我是「恨」了很久。現在的「獨
樂樂，不如眾樂樂」的美麗分享只是小試牛刀，但絕不是初試啼聲，因為黛黛早已
累積了不少實戰經驗。我知道，有一天，你必定會成為新一代的 cupcake 女神，擁有
自己的 boutique，跟大家分享精緻及具創意的甜點蛋糕。

Kelvin Tang

Founder of Treasure the Moment

一次江湖救急，黛黛成就了我人生中第一個親
手炮製的生日蛋糕，還給我家美人一次難忘的
超級驚喜！雖然之前已經品嚐過這位才女的出
品(實在是令人驚艷！)，但由跟她一起做蛋糕
的那天開始，我才真真正正感受到她對做蛋糕
的那份熱誠和執著。今次她更以世界風情為藍
本，自製特色 cupcake 攻陷你的芳心！就我本
人來說，一啖一個的 cupcake，想來都相當美
味……極之期待！撐！！！

姜皓文

藝人

黛黛是我香港小姐選美前一屆的得獎者，認識她不知不覺已有十年的時間。還記得當年我參加選美的時候，她常來探望我們，為我們這班小師妹們打氣加油，分享她選美時的經驗。那時候，黛黛給我的印象就是位非常貼心、溫暖、善良的女生。

選美之後雖然我們見面的時間不多，但她的溫暖依然，我們每次碰面總有聊不盡的話題。曾經有好幾次她在電視裏看到我的演出，都會發來短訊問候和支持。這位美貌與智慧並重的香港小姐就是這樣默默地、溫柔地關愛着他人，發放着正能量。

去年，我應邀作嘉賓出席黛黛的電台節目。她非常的體貼，在訪問的前一個星期就把訪問的內容、題目、時間和地點以中英文形式整齊、清楚地通過電郵發送給我，在電郵最後還窩心地補充說：「訪問前先給妳參考一下，一切以保護受訪嘉賓隱私、讓嘉賓舒服為前提。」我當時想，她應該擔心我不太會看中文字，所以花了不少時間寫了中英文對照翻譯版本給我。無論何時，她都很照顧別人感受，在她身邊都會感到輕鬆、自在。

她做事專注，細心關懷他人，對待美食態度認真，更有自己獨特的見解。雖然我踏入餐飲業時間不長，但在經營管理時都希望食物的質量和衛生能夠得到保證。而每當看到黛黛的美食作品時都不禁讓我讚嘆，這不僅僅是視覺上的享受，同時也打開了觀者的味蕾，做到了色香味俱全。

我很榮幸能夠為黛黛的新書作序，在這裏我非常誠意地向大家推薦這樣一位懂吃、愛吃、用心的美食家。謹在此祝福黛黛和各位讀者甜蜜、幸福、順心、快樂。

朱慧敏
2004年香港小姐亞軍及著名演員

目錄

一、踏上 Cupcake 之旅，Are You Ready？

>>>>>>>>>>>>>>>>>>>>>>>>>>>>>>

二、上路！把全世界的甜蜜裝進 Cupcake！

>>>>>>>>>>>>>>>>>>>>>>>>>>>>>>

香港 / 澳門 H.K/MACAU

台灣 TAIWAN

踏上Cupcake之旅

英國
比利時
意大利
法國
加拿大
美國
香港
澳門
日本
台灣
泰國
新加坡

Are You Ready?

常用工具 *Basic Equipments*

⑦ 電動打蛋器：手提烘焙攪拌器，可用於打蛋或打麵糰，可根據不同工序調整速度快慢。

⑨ 馬芬蛋糕烤盤：市場有多種不同材質可選，如傳統金屬烤盤、免粘烤盤或硅膠烤盤等。

① 量杯：烹飪用的量杯，1 杯 =250毫升。

⑪ 裱花唧嘴：用於蛋糕裝飾，可使用不同花型唧嘴做出不同形狀效果。

⑫ 剪刀：用於剪裁裱花袋。

⑧ 馬芬杯蛋糕紙托：一次性烘焙油紙蛋糕托，可配合派對主題選擇不同顏色和形狀。

⑤ 不銹鋼抹刀：適合一般需要用手攪拌的烘焙工序，或用於塗抹忌廉。

⑩ 一次性裱花袋：用於蛋糕裝飾，可用來唧牛油糖霜、忌廉芝士等。

④ 硅膠刮刀：耐高溫、防粘，適用於一般需要用手攪拌的烘焙工序。

❻ 攪拌碗：質料有玻璃、不銹鋼等，有不同尺寸可選擇，適合不同用途。

❶ **Measuring cups:** Standard one cup = 250ml in a glass measuring cup.

❷ **Electric kitchen scale:** For measuring dry or wet ingredients for baking in unit –grams.

❸ **Sifter / strainers:** For sifting dry ingredients, icing sugar, cocoa powder.

❹ **Rubber spatula:** For folding or scraping mixture from bowls and stirring hot liquid.

❺ **Metal spatula:** For normal mixing or cake frosting.

❻ **Mixing bowls:** Stainless steel or glass, need a variety of sizes for hand mixing.

❼ **Electric hand mixer:** Hand-held mixer with whisk attachment is most portable for mixing ingredients.

❽ **Cupcake liners:** For baking and part of decorating tool. Can buy in variety of colors and shapes.

❾ **Cupcake baking tray (muffin baking tray):** Comes in traditional stainless steel baking tray, non-stick aluminum baking tray, or silicone muffin baking tray.

❿ **Disposable piping bags:** A tool to apply frosting or piping on cupcakes for decoration purpose.

⓫ **Piping tips:** Tools to create different designs for cake decorating.

⓬ **Scissors:** For cutting the piping bags.

❷ 電子磅：家用電子磅，量度單位為克/安士。適用於量度烘焙用的乾、濕材料。

❸ 粉篩：用於篩撒麵粉、糖粉、可可粉等粉狀乾性烘焙材料。

蛋糕杯 *Baking Cups*

❶ 一次性防油紙杯：選用不同顏色、圖案和壓成不同形狀的烘焙用紙杯，可令杯子蛋糕擁有百變外形，為派對增添色彩和歡樂氣氛。一般烘焙用紙杯都可於超市或烘焙專門店找到，適用於多數馬芬蛋糕焗模。

❶ Paper liners: A part of cake decorating process with the use of paper liners to match party & color themes. You can find it in local supermarkets or bakery stores and are available in different sizes, shapes, solid colors or special printed designs that are suitable for standard muffin pans or cupcakes pans.

❷ 耐高溫硅膠焗模：耐用、耐高溫，可重覆多次使用的環保用具！硅膠是一種傳熱很快的材料，且有不粘的特性，很適合初学烘焙人士。

❷ Silicone liners: A good re-usable and environmental friendly choice for cake baking. They are durable, heat resistant and non-sticky when baking. Silicone is an excellent insulator that never gets hot while baking unlike traditional metal pans that will get very hot while in the oven.

常用食材 *Basic Ingredients*

❸ 發酵粉：蛋糕用發粉（膨脹劑），在烘焙加熱時會膨脹產生鬆軟效果。

❹ 蘇打粉：蛋糕用膨脹劑的一種，多用於較帶酸性的蛋糕，與果汁、酸乳酪、朱古力等材料搭配，有中和和鬆軟的效果。

❺ 忌廉芝士：用淡忌廉製作的新鮮軟質乾酪，具有濃郁的芝士味和特殊酸味，通常用於製作芝士類糕點。

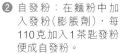

❶ 普通麵粉：低筋至中筋無發酵粉的麵粉，蛋白質含量低，筋度黏度較低，適用範圍廣。

❷ 自發粉：在麵粉中加入發粉（膨脹劑），每110克加入1茶匙發粉便成自發粉。

❻ 可可粉：又名谷咕粉，原味沒有添加糖粉，味道甘苦。可作蛋糕面裝飾。

❼ 淡忌廉：較厚身的糖流質忌廉，含脂肪最少35%或以上，打發後可作柔軟如雪糕般的唧花裝飾。

---

❶ **Plain Flour:** also known as all-purpose flour made from wheat.

❷ **Self-Raising Flour:** plain flour sifted with baking powder in the ratio of 1 cup flour to 10g baking powder.

❸ **Baking Powder:** a raising agent used mostly in cake baking and helps lighten the cake texture in baking.

❹ **Baking Soda:** also named bicarbonate of soda, a raising agent used in recipes with acidic ingredients such as lemon, chocolate, buttermilk, yogurt.

❺ **Cream Cheese:** a type of soft white smooth cheese, mild tasting with a high fat content.

❻ **Cocoa powder:** dried, unsweetened also known as cocoa.

❼ **Whipping cream:** thickened cream with minimum fat content of 35%, good for making a soft unsweetened light whipped cream frosting for cupcakes.

⑧ 酸奶：又名白脫牛奶
或酪奶，低脂，質地
似乳酪，味道微酸。
內含乳酸，可令蛋
糕、麵包更鬆軟。

⑨ 牛奶：一般飲用
鮮奶分為全脂、
低脂和脫脂三
種。市面上也有
售添加不同果汁
的果味奶。

⑩ 牛油：市面分無鹽
牛油和鹽味牛油兩
大類，內含動物性
脂肪。

⑪ 植物油：由不同
植物提煉，市面
上常見的植物油
有菜油、粟米油、
橄欖油等。

⑬ 食用色素：適合烘焙用之可食用
人工色素，食品添加劑的一種。
用於增強或改良食物顏色外觀。

牛奶朱古力　　　　　白朱古力　　　　　黑朱古力

⑫ 朱古力：由可可豆提煉而成，分為白朱古力、牛奶朱古力和黑朱古力三類。可溶
入烘焙蛋糕，或作朱古力味裝飾。

⑧ **Buttermilk:** a slightly sour thickened milk similar to yogurt. It's low in fat compared to fresh milk.

⑨ **Milk:** use full cream fresh milk or skimmed milk, it comes in different fruity flavors.

⑩ **Butter:** salted or unsalted. Use unsalted butter for regular cake baking. It contains all animal fats.

⑪ **Vegetable oil:** or salad oil, sourced from plant, it helps creates lighter sponge texture in cake baking.

⑫ **Chocolate:** available in milk, white, dark chocolate. Ideal for decorating or melt the chocolate into the cake batter for a luxury rich cocoa flavor.

⑬ **Food Colourings:** edible substance used to give colors to food, available in liquid, powdered or gel form.

⑭ 雞蛋：本書採用大size美國雞蛋，以重量約 50 克作為食譜標準。

⑮ 香精油：常用香草豆精油，用於增加香味。超市可買到不同果味的人工合成香精油。

⑯ 香料粉：烘焙用香料粉，常用於提升曲奇餅的味道。常見的有肉桂粉、薑粉、豆蔻粉等。

⑲ 糖霜粉：似粉狀的糖，呈白色粉末。市面購買的糖霜粉大多混入少許玉米澱粉作防潮之用。大多用於製作裱花牛油糖霜。

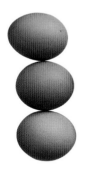

⑰ 白砂糖：白色、磨幼的食用砂糖，可增加甜味，亦可在打發雞蛋時起到穩定劑作用。

⑱ 紅糖：又名赤砂糖，帶有焦糖風味和不同深淺紅褐色。

⑳ 金黃糖漿：可用玉米糖漿、楓糖漿、蜜糖等代替。

⑭ **Eggs:** use large egg average weighting 50g each (standard US size L egg).

⑮ **Extracts/Essence:** non-alcoholic essences, which are available in different fruit flavours widely used in cake baking. Common extract used is vanilla extract made from vanilla beans.

⑯ **Spice:** grounded spices in powdered form, available in most supermarkets in a variety of different spices. Cinnamon, ginger, nutmeg is common used in festive cake or cookies recipes.

⑰ **Caster Sugar:** a soft finely granulated white table sugar.

⑱ **Brown Sugar:** comes in brown or dark brown sugar, a moist sugar with a rich full flavor & characteristic color.

⑲ **Icing Sugar:** confectioner sugar or powdered sugar, important base for buttercream or other frosting.

⑳ **Golden Syrup:** corn syrup, maple syrup or honey can be used as substitute.

單位及溫度換算
Conversion Chart & Oven Temperature

換算表 Conversion Chart

小貼示：將所有材料準確地用量杯或電子磅量好備用。材料如需用量杯、茶匙、湯匙等量度，注意先用湯匙底輕手抹平多餘份量。

Useful hints: all cup & spoon measurements need to be measured in level. The accurate way to measure all ingredients is to weigh them with a metric marked glass cup or electric kitchen scale.

乾材料 DRY ingredients		濕材料 LIQUID ingredients	
公制計量（克）Metric	英制計量（安）Imperial	公制計量（升）Metric	英制計量（安士）Imperial
15g	1/2 oz	30 ml	1 oz
30g	1 oz	60 ml	2 oz
60g	2 oz	100 ml	3 oz
90g	3 oz	125 ml	4 oz
125g	4 oz（1/4 lb）/ 磅	150 ml	5 oz
250g	8 oz（1/2 lb）/ 磅	250 ml	8 oz
375g	12 oz（3/4 lb）/ 磅	300 ml	10 oz
500g	16 oz（1 lb）/ 磅	500 ml	16 oz（1 lb）/ 磅
750g	24 oz（1 1/2 lbs）/ 磅	600 ml	20 oz（1 pint）/ 品脱
1 kg	32 oz（2 lbs）/ 磅	1000 ml	1 3/4 pint（1 litre）/ 公升

焗爐溫度 Oven Temperature

小貼示：注意不同牌子的家用焗爐溫度計算有分別，請先參考焗爐説明書。較小型家用焗爐可調較溫度低10度和烘焙時間縮短5-7分鐘左右。

Useful hints: Guide to conventional ovens (for fan-forced oven, check manufacturer's menu). For small home use oven, you can adjust the baking temperature to 10 degrees lower and shorten the baking time by 5-7 minutes.

攝氏 C-Celsius	華氏 F-Fahrenheit	焗爐溫度指數對照 Gas mark
120	250	1/2
150	275-300	1-2
160	325	3
180	350-375	4-5
200	400	6
220	425-450	7-8

基本蛋糕製作方法
Basic Cupcake Recipes

基本雲呢拿牛油蛋糕
Basic Vanilla Cupcakes

材料 Ingredients

45克無鹽牛油（室溫）	40毫升脫脂牛奶
70克白砂糖	20克植物油
1隻雞蛋	1/2茶匙香草香油
90克自發麵粉	

45g unsalted butter, at room temperature	90g self-raising flour, sifted
70g caster sugar	40ml skimmed milk
1 large egg	20g vegetable oil
	1/2 teaspoon vanilla essence

*本書提供的食譜均為供製作6個蛋糕的份量。
* All recipes in this book could be used to make 6 cupcakes.

步驟

1 準備
將焗爐預熱至攝氏170度。牛油在室溫中回軟。

2 打發牛油
加入白砂糖，和牛油一起打發，至顏色變淺，呈軟滑狀。

3 加入雞蛋
加入打散的蛋液，混合攪拌，直至和牛油融合。

4 加入過篩的粉類
將自發麵粉篩過，然後分3次加入。用切拌法快速拌勻。

5 加入其他材料
加入脫脂牛奶，攪拌均勻，再拌入植物油和香草香油。

6 入爐烘焙
在馬芬杯裏放入紙膜，倒入麵糊，約2/3滿。用170度烘烤20~30分鐘至呈金黃色。

Steps

1 Preheat the oven to 170 degrees Celsius.
2 Beat the butter and sugar until light yellow color and smooth.
3 Add in the egg and mix well.
4 Sift the self-raising flour. Add in the flour little by little to the batter.
5 Add in the milk little by little and then add in vegetable oil and the vanilla essence.
6 Pour into cupcake cups, 2/3 full. Bake for 20-30 minutes or until golden brown.

小貼士 Tips

如何判斷蛋糕是否熟透了？
在蛋糕中間插入牙籤，取出後如果牙籤有些濕濕的，代表蛋糕還未完成。相反，如果牙籤是乾身且有蛋糕碎黐住的話，表明蛋糕已經熟透了。

How to know if the cupcakes are ready?
Insert a toothpick in the middle of the cupcake and pull it out. If there are some watery fluid cupcake batter stuck on the toothpick, it means it is not ready. However, if the toothpick is dry with some cupcake pieces on it, then it is ready!

基本朱古力蛋糕
Basic Chocolate Cupcakes

材料 Ingredients

55克無鹽牛油（室溫）	90克麵粉
70克紅糖	60毫升酸奶（白脫牛奶）
1/2茶匙蘇打粉	1湯匙可可粉
1/8茶匙泡打粉（發酵粉）	30克黑朱古力（溶化）
1隻雞蛋	

55g unsalted butter, at room temperature	90g all purpose flour, sifted
70g brown sugar	60 ml buttermilk
1/2 teaspoon baking soda	1 tbsp cocoa powder (melted in 1 tbsp of hot water)
1/8 baking powder	30g dark chocolate chips melted and folded in batter
1 large egg	

步驟

1 準備
將焗爐預熱至攝氏170度。牛油在室溫中回軟。

2 打發牛油
加入紅糖，和牛油一起打發，直到顏色變淺，呈軟滑狀。

3 加入雞蛋
加入打散的蛋液，混合攪拌，直至和牛油融合。

4 篩入粉類
將麵粉、蘇打粉和發酵粉混合過篩，然後分3次加入。

5 加入牛奶、可可粉
依次加入白脫牛奶和可可粉醬。

6 加入朱古力
排入溶化的黑朱古力。

7 入爐烘焙
在馬芬杯裏放入紙膜，倒入麵糊，約2/3滿。用170度烘烤20~30分鐘至呈金黃色。

Steps

1 Preheat the oven to 170 degrees Celsius.

2 Beat the butter and sugar until light yellow in color and smooth.

3 Add in the egg and mix well.

4 Sift the flour, baking soda and baking powder together. Add in the flour little by little to the batter.

5 Add in the buttermilk and cocoa paste little by little.

6 Stir in the melted chocolate.

7 Pour into cupcake cups, 2/3 full. Bake for 20-30 minutes or until golden brown.

基本糖霜製作方法
Basic Icing Recipes

基本牛油糖霜
Basic Buttercream

材料 Ingredients

100克無鹽牛油（室溫）	2~3茶匙食用凍水/脱脂牛奶	100g unsalted butter, at room temperature	2-3 teaspoons of skimmed milk or cold water
200克糖霜粉（已篩）	1茶匙雲呢拿香油	200g icing sugar (sifted)	1 teaspoon vanilla essence

步驟

1 打發牛油和糖粉
牛油在室溫中回軟，用攪拌器打至軟滑，慢慢篩入糖粉，繼續攪拌。

2 加入液體
慢慢加入飲用凍水或牛奶，再加入雲呢拿香油。

3 繼續打發
繼續打發至鬆軟滑身。

Steps

1 Beat the butter at medium high speed until soft and creamy. Add in the sifted icing sugar little by little, beat at high speed until all blend in.

2 Add in the milk or cold water one tablespoon at a time.

3 Beat at high speed until icing is smooth and fluffy.

小貼士 Tips

1 將牛油糖霜放入密封容器，可於雪柜冷藏格儲存4-5天。

2 唧花前，應先將牛油糖霜放至室溫，才開始唧上蛋糕面。

1 Store in air tight container inside fridge for 4-5 days.

2 Bring to room temp before frosting on cakes.

蛋白霜
Meringue Frosting

此食譜蛋白霜質感輕盈，
入口感軟滑如雪糕般！

材料 Ingredients

糖漿：
165克白砂糖
15克(1湯匙)粟米糖漿
30克飲用凍水

蛋白霜：
99克(3隻大Size雞蛋) 蛋白
15克(6茶匙)白砂糖

Syrup:
165g granulated sugar
15g (1 tbsp) corn syrup
30g water

Meringue:
99g (3 large sz) egg white
15g (6tsp) granulated sugar

步驟

1 煮沸糖水
在小鍋中放入水、粟米糖漿和白砂糖，煮沸至溫度240華氏，約攝氏110度。

2 打發蛋白
將蛋白放入另一打蛋盆中，用攪拌器中速打發出 泡沫，分多次逐少加入砂糖，

繼續用中高速打發蛋白到企身。

3 混合
將煮沸糖水慢慢倒入蛋白霜內，繼續用高速打發蛋白霜至光亮，鬆軟順滑，待蛋白霜降回室溫才可放入唧袋開始唧花步驟。

Steps

1 Bring icing sugar, corn syrup, water in a saucepan to broil at 240F (110 degrees Celsius).

2 In another bowl, beat egg whites & white sugar to firm peaks.

3 Pour in hot syrup in a slow stream down from side of bowl into egg white, and beat on high speed until meringue looks fluffy and glossy, cool down before frosting on cupcakes.

法式蛋白奶油霜
French Meringue Buttercream Frosting

材料 Ingredients

40克（1隻加大size雞蛋）蛋白　110克無鹽牛油（室溫）
45克白砂糖

40g (1 xl sz) egg white　110g unsalted butter
45g sugar　(softened in room temp)

步驟

1 打發蛋白
將蛋白放入打蛋盆，用攪拌器中速打發蛋白成　泡沫狀。之後逐少加入砂糖，繼續高速打發至企身。

2 加入牛油
開始加入室溫牛油，要分多次拌勻入蛋白霜內約30克（2湯匙量）。

3 繼續打發至濃稠
攪拌器高速繼續打發至奶油霜變成濃稠順滑。

Steps

1 In a bowl, beat egg whites to foamy. Add in sugar and beat to stiff peaks.

2 Add in butter (soft room temp) 2 tablespoons each time into the meringue.

3 Continue to beat at high speed until buttercream starts to thicken and becomes smooth.

小貼士 Tips

當加入軟牛油粒時，蛋白霜與牛油會形成分離狀態，這是正常的。繼續打發約5-10分鐘後，蛋白奶油會自然混勻。

When you first add in the butter, meringue will break down and curd. Keep beating in butter at high speed and buttercream will form after 10-15 minutes time.

忌廉芝士糖霜
Cream Cheese Frosting

材料 Ingredients

100克忌廉芝士（室溫放軟）

100克糖霜粉(已篩)
1/4 茶匙香草香油(可選)

100g cream cheese (softened in room)
100g icing sugar (sifted)

1/4 teaspoon vanilla extract (optional)

步驟

1 打發忌廉芝士
將室溫忌廉芝士放入打蛋盆中，用攪拌器中速打至鬆軟。

2 加入糖粉
分6-7次逐少加入糖粉。

3 繼續打發
用電動攪拌器繼續打發至鬆軟順滑。

Steps

1 Beat the cream cheese until soft and creamy.

2 Add in the sifted icing sugar little by little.

3 Beat at high speed until icing is smooth and fluffy.

朱古力醬
Chocolate Ganache

材料 Ingredients

100克淡忌廉
200克60%黑朱古力

* 朱古力對淡忌廉比例2:1
（若用於淋面可調整至1:1
或1:1.5）

whipping cream 100g
60% dark chocolate 200g

* Ratio – chocolate 2 : heavy
whipping cream 1 (adjust to 1:1/
1:1.5 if you need a more softer
creamy consistency)

步驟

1 加熱淡忌廉
將淡忌廉倒入小碗內，放入微波爐加熱2-3
分鐘至邊緣冒泡微沸狀態。

2 融化朱古力
朱古力置於另一碗中，倒入熱忌廉，靜待
2-3分鐘至朱古力開始融化。

3 混合
用抹刀輕輕將朱古力和忌廉攪拌至完全均
勻，呈光亮軟滑狀。

4 完成
完成後放入唧袋，或可用抹刀將朱古力醬抹
平於蛋糕面。

Steps

1 Microwave whipping cream for
2-3 minutes until hot but not
boiling.

2 Pour hot cream into chocolate
buttons until it melts.

3 Stir with spatula until it all blend
in to form a smooth paste.

4 Place into piping bag or pour
onto the cupcakes.

小貼士
Tips

1 如果需要平鏡面淋醬
效果，可立即淋上蛋
糕面待涼。

2 如需要做唧花塑形，
將朱古力醬蓋一張保
鮮膜放入雪櫃約半小
時（約10分鐘可查看
朱古力醬是否達到理
想軟硬度，若冷藏太
久會變硬，不適合唧
花）。

1 Use immediately and pour onto
the cupcakes if you want a liquid
consistency.

2 However, if you need a firmer
consistency for piping or frosting,
cover the chocolate with plastic
wrap and chill it in the fridge for 30
minutes. (Check every 10 minutes
to see if the Ganache has already
reached the right consistency for use
or else it will be become too hard.)

基本裝飾技巧
Basic Decoration Tips

預備裱花袋 Fill a Piping Bag

工具 Tools

一次性裱花袋	disposable plastic piping bag (or you can use zip-lock bag at home)
裱花唧嘴	piping tip
攪拌抹刀	spatula
高身玻璃杯	a tall glass
剪刀	scissors
橡皮筋	rubber band

步驟

1 準備裱花袋
預備一個一次性裱花袋，在袋尾尖處剪1個約1吋至1.5吋的開口(約裱花嘴長度一半)。

2 放入唧嘴
將唧嘴放入裱花袋內。唧嘴較窄的一端要突出於裱花袋口外。

3 套入玻璃杯
將裱花袋套入1個高身玻璃杯，裱花袋口向下包住玻璃杯口。

4 放入牛油糖霜
用抹刀將預備好的牛油糖霜放入袋內，用手將牛油糖霜推至袋尖，並推出袋內空氣。

5 密封
用手將裱花袋口按實，用橡皮筋將裱花袋口密封。

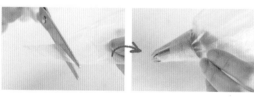

Steps

1 Get a disposable piping bag, use scissors to cut out the narrow tip off with 1 to 1.5 inches (depending on the tip size).

2 Select a piping tip to be used and drop it into your piping bag. Fit the tip to the narrow end of the piping bag.

3 Put the bag into a tall glass and fold the outside of the piping bag over the lip.

4 Use a spatula and fill in frosting into piping bag. Use your hand to try scraping down the icing to the bottom of the tip to get rid of all the air bubbles.

5 Lift the bag out of the glass, use your hand to twist the top of the piping bag and tie it up with a rubber band. This can avoid icing spread out during piping.

調色技巧 *Make Color Icing*

材料 Ingredients

牛油糖霜	buttercream icing
調色杯	mixing cups
食用色素	edible food color
牙簽	toothpicks
抹刀	spatula

步驟

1 打發牛油
先打發好適量白色牛油糖霜備用。

2 混入色素
用牙簽挑出少量色素，再混入牛油糖霜中。

3 拌勻
用手抹刀以打圈式將顏色拌勻於糖霜內。

4 重複步驟
如要將顏色再調深，可用一支新牙簽挑出顏色重複步驟(2)(3)。

Steps

1 Start with basic original white buttercream icing.

2 Insert toothpick into the food coloring bottle to pick up a small amount and swirl into icing.

3 Use a spatula to blend in color to icing.

4 Use a new clean toothpick to add more color if the color is not enough.

唧嘴使用技巧 *Piping Tips*

常見唧嘴

① 多瓣花齒唧嘴
可唧出較動感的波浪紋線條效果。

② 開縫星齒唧嘴
可唧出較細緻的扭繩紋或立體多層雪糕
筒效果。

③ 圓口裱花唧嘴
可唧出蜂巢圈效果或用特小圓口裱花唧
嘴作寫字劃線裝飾。

④ 開縫法式裱花唧嘴
可唧出蓬松狀的花紋。

Helpful hints to choose a tip

① Use a Closed Star Piping Tip
If you want a shorter lower frosting or more "ruffly" look swirls.

② Use a Open Star Piping Tip
If you want a taller, ice cream frosting or a thicker rope look.

③ Use a Round Piping Tip
If you need to draw a line or shapes on cupcake design.

④ Use a French Star Piping Tip
If you want to pipe fluffy curls, or shell patterns.

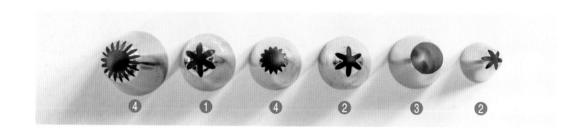

④　　**①**　　**④**　　**②**　　**③**　　**②**

漩渦狀裝飾 *The Perfect Swirl*

工具

多瓣花齒唧嘴（使用不同大小的唧嘴及花齒可做出不同效果）

Tool

A medium size Closed Star Piping Tip (or use different tips using same technique)

步驟

1 唧袋垂直於蛋糕平面
注意唧袋位置要垂直於蛋糕平面成90度角。

2 控制擠壓力度
用拇指、食指按緊唧袋頂部，其餘手指平均按緊唧袋，以控制擠出的糖霜量。

3 唧出第一層
將手按緊唧袋頂部，於蛋糕中心位置唧出適量糖霜，再轉一小圈。繼續由內圈擴展至蛋糕外圈形成第一層。

4 重複
用相同唧法再疊上第二、第三層。當到達最頂層，放鬆手指，停止按壓唧袋及向上抽離。

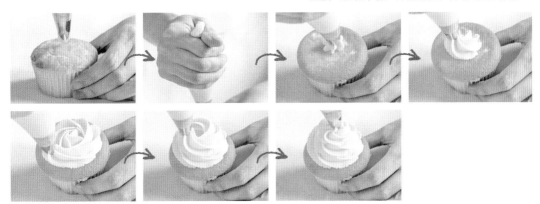

Steps

1 Hold piping bag in vertical up right position to your cupcake.

2 Apply even pressure to your frosting bag.

3 Start piping out from the center of cupcake. Pipe a large circle from centre out and around the outer-edge of cupcake.

4 Continue piping smaller circles on top of one another. Release pressure (stop squeezing the bag) then lift up.

星形花裝飾 *Star Drop Flower*

工具
小號或中號開縫法式裱花唧嘴

Tool
A small to medium French Star Piping Tip

步驟

1 **唧袋垂直於蛋糕平面**
注意唧袋位置要垂直於蛋糕平面成90度角。

2 **控制擠壓力度**
將拇指食指按緊唧袋頂部，其餘手指平均按緊唧袋，以控制按壓出糖霜多少量。

3 **唧出星形花**
手指平均用力，按緊唧袋頂部，唧出適量糖霜，於蛋糕沿外圍完成一圈星形花。當完成每一粒星形花，放鬆手指，停止按壓唧袋及向上抽離。

4 **重複**
重複以上步驟，直至星形花平均遮蓋整個蛋糕面。

Steps

1 Hold the piping bag in upright position, then Apply pressure and start at the outer edge of the cupcake.

2 Apply even pressure to your frosting bag.

3 Let the icing come out quickly, move tip upwards and stop the pressure.

4 Repeat the process for the entire cupcake and finish the last one for the centre.

蓬鬆球裝飾 *Fluffy Curl*

工具
中號或大號開縫法式裱花唧嘴

Tool
A medium or big French Star Piping Tip

步驟

1 唧袋垂直於蛋糕平面
注意唧袋位置要垂直於蛋糕平面成90度角。

2 控制擠壓力度
將拇指、食指按緊唧袋頂部，其餘手指平均按緊唧袋，控制按壓出糖霜多少量。

3 唧出蓬鬆球
手指平均用力，按緊唧袋頂部，於蛋糕中心位置唧出適量糖霜。繼續輕力按壓唧袋，並將手慢慢提升唧出糖霜，令中心疊起形成一個微捲小球。

4 完成
當唧出至需要高度，放鬆手指，停止按壓唧袋及向上抽離。

Steps

1 Hold the piping bag vertically and apply the pressure from the center of the cupcake.

2 Apply even pressure to your frosting bag.

3 Apply more pressure, let the icing come out and raise the tip slightly in a slow motion.

4 Stop the pressure when it reaches the tall end and lift the tip straight up.

33

玫瑰花裝飾 *Rose*

工具

中號或大號開縫星齒裱花唧嘴

Tool

A medium or large Open Star Piping Tip

步驟

1 唧袋垂直於蛋糕平面
注意唧袋位置要垂直於蛋糕平面成90度角。

2 控制擠壓力度
將拇指、食指按緊唧袋頂部，其餘手指平均按緊唧袋，控制按壓出糖霜多少量。

3 唧出玫瑰花
手指平均用力，按緊唧袋頂部，唧出適量糖霜於蛋糕中心位置轉一小圈。繼續唧出糖霜，沿內圈向外繞出一至兩圈，至到離蛋糕紙托邊約5mm-8mm。

4 完成
當完成最後外圈，放鬆手指，停止按壓唧袋及向側面抽離。

Steps

1 Hold the piping bag in a vertical up right position.

2 Apply a slight pressure and make a dollop in center.

3 Without releasing the pressure, slowly move tip out from center and start make circle around center dollop.

Complete with one to two circles (depending on the size of the star piping tip you use) around the center.

4 Finish by sliding the tip down to the outer-edge of cupcake and release the pressure and pull the tip off.

小花群裝飾 *Blossoms*

工具
小號或中號開縫星齒裱花唧嘴

Tool
A small or medium Open Star Piping Tip

步驟

1 唧袋垂直於蛋糕平面
   注意唧袋位置要垂直於蛋糕平面成90度角。

2 控制擠壓力度
   將拇指食指按緊唧袋頂部，其餘手指平均按緊唧袋，控制按壓出糖霜多少量。

3 唧出小花
   手指平均用力，按緊唧袋頂部，唧出適量糖霜，沿蛋糕外圍成一粒粒花形效果。每完成一粒小花，放鬆手指，停止按壓唧袋及向上抽離。

4 重複
   重複以上步驟至小花群均勻遮蓋整個蛋糕面。當完成最外圈，放鬆手指，停止按壓唧袋及向側面抽離。

Steps

1 Hold the piping bag in an up-right position.

2 Start piping out the blossom from the outer circle of the cupcake.

3 Apply pressure to let the icing come out to form a blossom, then quickly lift the tip up and release pressure on the bag.

4 Repeat that same motion around the entire cupcake. Finish the last blossom in the centre.

盤繞線圈裝飾 *Coil*

工具
大圓口裱花唧嘴

Tool
A large Round Piping Tip.

步驟

1 唧袋垂直於蛋糕中心
注意唧袋位置要垂直於蛋糕平面成90度角。

2 唧出線圈
手指平均用力，按緊唧袋頂部，唧出適量糖霜，螺旋圍繞中心點，並繼續圍繞著蛋糕的邊緣，唧一圈。

3 唧出第二圈
回到起點，繼續唧第二圈。輕輕重疊於第一圈之上，做成好像線圈的樣子。

4 重複
重複同樣的步驟，最後放鬆手指，向上拉起，形成一個尖嘴。

Steps

1 Start at the center of your cupcake. Hold the piping bag in vertical up right position.

2 Apply even pressure. Pipe a spiral around center point and continue to make a circle around the border of the cupcake.

3 When you get to the starting point, continue piping the second circle slightly overlapping the first circle to create a coil look.

4 Repeat the same process, continue to swirl the top, release pressure and pull tip up to form a soft peak.

簡易蛋糕裝飾 *Simple decoration*

利用不同形狀、顏色的糖果可將普通的杯子蛋糕裝點出不同的風格。

It is a simple, unique way to use candies, chocolates, sugar beads in all shapes and sizes as decorative toppers to create your own frosted cupcakes with an individual design.

1 派對紙製裝飾
Party paper toppers

2 彩色飲管
Drinking straws

3 紅色裝飾彩砂糖
Red color sanded sugar

7 彩色圓形小糖珠
Round sugar candies

4 皇室糖霜小花
Sugar flowers

5 薄荷朱古力粒
Mint flavor chocolate chips

6 黑朱古力薄片
Dark chocolate flakes

8 綠色裝飾彩砂糖
Green color sanded sugar

9 彩色朱古力豆
Chocolate beans

10 黑朱古力粒
Dark chocolate chips

11 彩色迷你糖珠
Rainbow sugar beads

12 白朱古力粒
White chocolate buttons

13 迷你彩色朱古力豆
Mini chocolate beans

14 草莓味花瓣朱古力片
Strawberry flavored white chocolate curls

15 食用裝飾銀珠糖
Silver sugar beads

16 棉花糖
Marshmallows

17 迷你黑朱古力粒
Mini dark chocolate chips

18 天然免焗碎杏仁粒
Dried chopped almond

19 花瓣白朱古力片
White chocolate curls

20 糖衣軟糖
Sugar coated jelly beans

21 黃色裝飾彩砂糖
Yellow color sanded sugar

22 皇室糖霜紅蘿蔔裝飾
Sugar carrots toppers

上路！
把全世界的甜蜜裝進
Cupcake！

香港·澳門

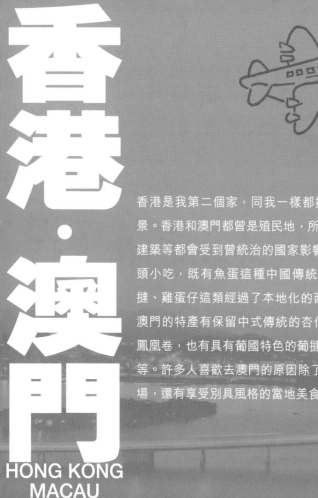

香港是我第二個家，同我一樣都擁有中西文化兩種背景。香港和澳門都曾是殖民地，所以好多文化、飲食、建築等都會受到曾統治的國家影響。比如香港的街頭小吃，既有魚蛋這種中國傳統美食，也有蛋撻、雞蛋仔這類經過了本地化的西式食物；而澳門的特產有保留中式傳統的杏仁餅、蛋捲、鳳凰捲，也有具有葡國特色的葡撻、木糠布甸等。許多人喜歡去澳門的原因除了光顧賭場，還有享受別具風格的當地美食。

HONG KONG
MACAU

好友車車
的美食印象

放假不想留在香港，又嫌搭飛機路程太遠。去澳門兩日一夜是一個不二之選。澳門又是一個食都，基本上一日 24 小時，你可以食到四、五餐。雖然香港現在也有某些杏仁餅或豬肉乾專門分店，但香港遊客始終去到澳門都一定要買些回港才覺得不枉此行，或者這就是「人家的飯特別好吃」的心理。至於我們的香港，其實都有好多特色小食吸引外國人。例如咖哩魚蛋、雞蛋仔、砵仔糕、雞尾包、菠蘿油等，這些小食的滋味真的只有香港才能品嚐到。所以，住在香港這個美食天堂確實很有口福。

芝麻蛋糕

Sesame Cupcakes

芝麻糊是一種中國傳統甜點，香港很多甜品店都
可以吃到。但是，有時覺得芝麻糊比較甜膩，
如果吃一整碗的話，會感到很飽滯。
這款黑芝麻cupcake的內陷和糖霜都加了芝麻糊，
充滿濃香的芝麻滋味，又避免了飽腹感，
適合喜愛芝麻口味的你！

材料 Ingredients

蛋糕
55克無鹽牛油（室溫）
50克白砂糖
1隻雞蛋
100克自發粉
1.5湯匙芝麻粉
40毫升脫脂牛奶
25克植物油
10克烤黑芝麻粒
1滴黑色食用色素

芝麻牛油糖霜
100克無鹽牛油（室溫）
200克糖霜粉
2-3茶匙脫脂牛奶
2湯匙芝麻粉
1滴黑色食用色素

裝飾
1個花瓣唧嘴
1個唧袋
適量烤黑芝麻粒

For Cupcake
55g unsalted butter, at room temperature
50g caster sugar
1 large egg
100g self-raising flour
1.5 tablespoon instant sesame dessert powder
40ml skimmed milk
25g vegetable oil
10g toasted black sesame
1 drop of black food coloring

For Sesame Buttercream
100g unsalted butter, at room temperature
200g icing sugar
2-3 teaspoons skimmed milk
2 tablespoons instant sesame dessert powder
1 drop of black food coloring

For Decoration
Petal Piping Tip – ruffle flower
1 piping bag
toasted black sesame sprinkle on top

Cupcake Steps

1 Preheat the oven to 170 degrees Celsius.

2 Beat the butter and sugar until smooth.

3 Add in the egg and mix well.

4 Sift the self-raising flour. Add in the flour and sesame dessert powder little by little to the batter.

5 Add in the milk and vegetable oil little by little. Stir in the toasted black sesame. Add in a drop of black food coloring.

6 Pour into cupcake cups, 2/3 full. Bake for 20-30 minutes or until golden brown.

7 Transfer to cooling rack. Cool down completely before frosting.

Sesame Buttercream Steps

1 Beat the unsalted butter until soft. Sift the icing sugar and add in little by little.

2 Add in the milk one teaspoon at a time. Mix in the instant sesame dessert powder. Add in the black food coloring.

3 Beat until smooth and creamy.

Decoration Steps

1 Place the Petal Piping Tip into a piping bag. Place the Sesame Buttercream into the piping bag and pipe onto each cupcake. Pipe out with tip position at 45 degree angle to cupcake, apply with steady pressure, pipe out icing on cupcake in an up & down motion going around to form a ruffle flower.

2 Sprinkle toasted black sesame in the middle (decorate as core of the flower).

製作蛋糕

1 準備
將焗爐預熱至攝氏170度。牛油置於室溫回軟。

2 打發牛油
加入白砂糖，和牛油一起打發，直到顏色變淺，呈軟滑狀。

3 加入雞蛋
加打散的蛋液，混合攪拌，直至和牛油融合。

4 篩入粉類
將自發麵粉過篩，然後分3次加入。用切拌法快速拌勻。

5 加入其他材料
依次加入芝麻粉、植物油、烤黑芝麻和1滴黑色食用色素。

6 入爐烘焙
在馬芬杯裏放入紙膜，倒入麵糊，約2/3滿。用170度烘烤20~30分鐘或至表面呈金黃色。

7 冷卻
完成後，轉移到散熱架上，等蛋糕完全冷卻後才裝飾。

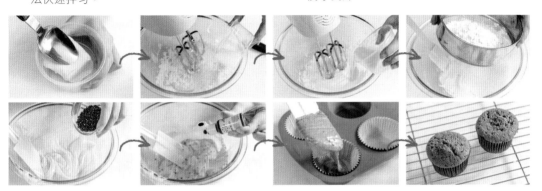

製作芝麻牛油糖霜

打發牛油
先將室溫牛油用電動攪拌器打至軟身，分5-6次加入糖霜粉。

加入其他材料
逐茶匙加入脫脂牛奶。加入芝麻糊甜品粉，然後拌勻。加入1滴黑色食用色素。

攪拌均勻
打發所有材料至鬆軟滑身。

裝飾步驟

使用花瓣唧嘴
拿1個唧袋，放入花瓣唧嘴。將芝麻牛油糖霜加入唧袋，往下推實，紮緊袋口，在蛋糕表面唧上花紋。開始唧花時，唧嘴要和蛋糕成45度角，輕輕唧出糖霜，手微向前後推動呈小波浪動作。先完成外圈圍邊再繼續完成內層。

撒上芝麻粒
最後撒上烤黑芝麻粒作裝飾。

完成！

菠蘿包蛋糕
Pineapple Bun Cupcakes

菠蘿包是香港的特產，誕生於60年代，是港式茶餐廳的經典
食物！只要加一塊牛油就變身美味的菠蘿油，
再配搭一杯又香又滑的奶茶，就是香港人最佳的「三點三」。
而 cupcake 是外國人的小食，同菠蘿油一樣，
是下午茶的主角。所以我做了這個 fusion 版的菠蘿包 cupcake：
一次吃兩款下午茶點，是不是很過癮呢？

材料 Ingredients

蛋糕

1份基本雲呢拿牛油蛋糕麵糊

菠蘿皮

62.5克低筋麵粉或普通麵粉
27.5克白砂糖
20克無鹽牛油（室溫）
3.5克奶粉
1隻雞蛋黃（1/2個用於菠蘿脆
　皮，1/2個用於蛋糕表層）
1/2湯匙淡奶
1/2茶匙煉奶
1/2茶匙發酵粉
1/4茶匙蘇打粉

For Cupcake

1 portion of Basic Vanilla Cupcake
　Batter

For Pineapple-bun Topping

62.5g cake flour or plain flour
27.5g caster sugar
20g unsalted butter (in room
　temperature)
3.5g milk powder
1 egg yolk (1/2 for the bun topping,
　1/2 for brushing)
1/2 tablespoon evaporated milk
1/2 teaspoon condensed milk
1/2 teaspoon baking powder
1/4 baking soda

製作蛋糕

準備麵糊

在馬芬杯裏放入紙膜，倒入雲呢拿牛油蛋糕麵糊，約2/3滿。

製作菠蘿皮

打發牛油

牛油置於室溫回軟，加入糖，打至滑身。 →

加入奶類和蛋黃

依次加入奶粉、半隻蛋黃、淡奶、煉奶，攪拌均勻。

篩入粉類

將低筋麵粉（或普通麵粉）、發酵粉、蘇打粉混合過篩，加入其中，攪拌成麵糰。 →

冷藏麵糰

將麵糰做成長圓筒形，包上保鮮紙，放進雪櫃冷藏30-40分鐘。

壓成圓片狀

從雪櫃拿出麵糰分六等份，再用手搓成圓球狀，然後壓平至圓片狀（直徑約2吋）。

用小刀劃出花紋

用小刀輕劃菠蘿皮表面，劃出深約1毫米的格子紋路，注意不要切斷脆皮。

烘焙

將菠蘿皮放在蛋糕糊的上面，在表面掃上蛋黃汁，用170度烘烤20~30分鐘或至呈金黃色。

裝飾步驟（可選）

把蛋糕切成一半，中間夾放一塊凍牛油厚片。

Pineapple Cupcake Steps

Prepare 1 portion of Basic Vanilla Cupcake Batter. Pour into cupcake cups, 2/3 full.

Pineapple-bun Topping Steps

1　Beat the butter until smooth. Add in the sugar until white and fluffy.

2　Add in milk powder, 1/2 egg yolk, evaporated milk, condensed milk.

3　Sift the cake flour, baking powder and baking soda and add into mixture.

4　Wrap up the dough into a long cylinder shape and refrigerate it for 30-40 mins, use cling wrap to wrap as its very sticky and wet.

5　Divide the dough into 6 even portions and roll each into a round ball. Use your palm to push each ball into a flat round disc shape.

6　Place and centre on top of each cupcake. Use a knife to draw diamond grids on top.

7　Brushed with the remaining egg yolk. Bake together with the cupcakes for 20-30 minutes or until golden brown.

For Decoration (optional)

Cut the cake into half and place in the cold butter slice.

杏仁蛋糕
Almond Cookies Cupcakes

如果要選一個代表澳門的食品，應該非杏仁餅莫屬。
杏仁餅本身較乾身一點，有很香的杏仁味同杏仁粒。
這個杏仁餅cupcake不但杏仁味十足，而且口感綿軟，
令你可以嚐到杏仁餅的另一種風格。

杏仁味
十足！

製作蛋糕

1 加入杏仁香油、碎粒
預備好一份的基本雲呢拿牛油蛋糕糊。放入杏仁香油、杏仁碎粒。

2 入爐烘焙
在馬芬杯裏放入紙膜，倒入麵糊，約 2/3 滿。用 170 度烘烤 20-30 分鐘或至蛋糕表面呈金黃色。

3 冷卻
完成後，轉移到散熱架上，等蛋糕完全冷卻後才裝飾。

加入杏仁香油、碎粒

製作杏仁牛油糖霜

加入杏仁香油
將杏仁香油加入預備好的牛油糖霜裏，攪拌均勻。

打發糖霜
用電動攪拌器打發至滑身。

裝飾步驟

使用開縫星齒唧嘴
拿 1 個唧袋，放入開縫星齒唧嘴。將杏仁牛油糖霜加入唧袋，往下推實，紮緊袋口，在蛋糕表面唧上花紋。

放上杏仁曲奇和杏仁碎粒
放上杏仁曲奇，再在蛋糕面上灑一些杏仁碎粒。

材料 Ingredients

蛋糕
1 份基本雲呢拿牛油蛋糕麵糊
1 茶匙杏仁香油
15 克杏仁碎粒

杏仁牛油糖霜
一份基本牛油糖霜
1/2 茶匙杏仁香油

裝飾
1 個開縫星齒唧嘴
1 個唧袋
適量杏仁碎粒
適量杏仁曲奇

For Cupcakes
1 portion of Basic Vanilla Cupcake Batter
1 teaspoon almond extract
15g almond bits

For Almond Buttercream
1 portion of Basic Buttercream
1/2 teaspoon almond extract

For Decoration
Open Star Piping Tip
1 piping bag
almond bits for decoration
almond cookies

Cupcake Steps

1 Add the almond essence into 1 portion of Basic Vanilla Cupcake Batter. Stir in the almond bits.

2 Pour into cupcake cups, 2/3 full. Bake for 20-30 minutes or until golden brown.

3 Transfer to cooling rack. Cool down completely before frosting.

Almond Buttercream Steps

1 Add the almond extract into 1 portion of Basic Buttercream.

2 Mix until well blended.

Decoration Steps

1 Place the Open Star Piping Tip into a piping bag. Place the Almond Buttercream into the piping bag and pipe onto each cupcake.

2 Decorate with almond cookies. Sprinkle almond bits on top.

木糠布甸蛋糕

Serrandura Cupcakes

木糠布甸是澳門別具特色的葡式甜品。
不同的餐廳會有不同的做法。木糠是指餅乾碎，
而布甸是由忌廉等材料組成的。我每次去澳門都會吃這個甜
品，但是每次都覺得好像欠缺一些東西，口感太 creamy，
所以我加了蛋糕去做這個木糠布甸 cupcake，
來填補我覺得欠缺的東西。

材料 Ingredients

蛋糕
1份基本雲呢拿牛油蛋糕麵糊

裝飾
1個開縫星齒唧嘴
1個唧袋
250毫升淡忌廉
2湯匙煉奶
2塊消化餅乾(壓碎)

For the cupcake
1 portion of Basic Vanilla Cupcake Batter

For Decoration
Open Star Piping Tip
1 piping bag
250ml whipped cream
2 tablespoons condensed milk
2 pieces of digestive biscuits (crumbs)

製作步驟

1 製作基本牛油蛋糕
把準備好的麵糊倒入馬芬杯,約2/3滿。用170度烘烤20~30分鐘或至呈金黃色。

2 填入煉奶
在每一個蛋糕中間挖一個小孔,在小孔中間填滿煉奶。

3 打發淡忌廉
將淡忌廉打至企身,加入2湯匙煉奶,拌勻。

4 唧上花紋
拿1個唧袋,放入開縫星齒唧嘴。將已打起的鮮忌廉放入唧袋,往下推實,紮緊袋口,在蛋糕表面唧上花紋。

5 撒上消化餅乾碎
最後,撒一些壓碎的消化餅乾作木糠裝飾。

Steps

1 Pour 1 portion of Basic Vanilla Cupcake Batter into 6 cupcake cups, 2/3 full. Bake for 20-30 minutes or until golden brown.

2 Cut a hole in the middle of the cupcakes. Fill with condensed milk.

3 Whip the cream to peak. Fold in the condensed milk.

4 Place the Open Star Piping Tip into a piping bag. Place the whipped cream into the piping bag and pipe onto each cupcake.

5 Sprinkled with digestive biscuit crumbs.

台灣
TAIWAN

台灣由於地處熱帶及亞熱帶氣候區之交界，自然景觀與生態資源相當豐富。人口以原住民族和漢族為兩大民族，或分為原住民族、河洛人、客家人、外省人、新住民等 5 大族群。因為有那麼多不同民族令到食物也好多元化，也使各族群傳統的食物做法不會失傳。每次到台灣必去逛夜市，各種當地美食讓人吃到停不了嘴！鳳梨酥、太陽餅、珍珠奶茶、炸雞、肉燥飯、三杯雞等等，都是台灣的招牌美食，絕對值得一試！

好友車車
的美食印象

問十個香港人「你喜歡台灣嗎？」，應該有九個都會答你「喜歡」。最後一個應該會答你「很喜歡」。原因有太多，為美食、為台妹、為夜蒲……這個地方確實令人感到快樂，好像人人都是禮貌大使，讓人感到好溫暖，難怪許多香港人去了就不想走！我更是瘋狂愛上台灣，試過一年去六、七次，差點被誤會有一個台灣男友。

如果要用一種食物去代表台灣……真的很難選擇，因為人人都知道台灣美食遍地都是。但如果要將一種台灣美食與 cupcake 結合，我想鳳梨酥 cupcake 應該好好味！再配上一杯黑糖薑母茶或珍珠奶茶，簡直 perfect! 但是，我本人比較古靈精怪。黛黛，下次可唔可以試整滷肉飯 cupcake 或蚵仔煎 cupcake 呢？哈哈！

車車

鳳梨蛋糕
Pineapple Cupcakes

鳳梨酥在台灣已經有太多不同的品牌，
味道更是五花八門，已經進化到有草莓味或芒果味，
甚至有些加入蛋黃。我這款鳳梨味道的cupcake，
不但有新鮮鳳梨粒和鳳梨餡，
就連蛋糕的表面都做出鳳梨的樣子。
不知道這個鳳梨cupcake將來會不會取代鳳梨酥，
成為另一種熱門手信呢？

材料 Ingredients

蛋糕

55克無鹽牛油（室溫）
50克白砂糖
1隻雞蛋
100克自發粉
40毫升脫脂牛奶
20克植物油
1/2茶匙鳳梨香油
30克罐頭或新鮮菠蘿切粒

鳳梨牛油糖霜

一份基本牛油糖霜
1/2茶匙鳳梨香油
黃色食用色素
綠色食用色素

裝飾

開縫星齒唧嘴
葉形唧嘴
兩個唧袋
鳳梨餡（即食餡料，烘焙店可買）

For Cupcake

55g unsalted butter, at room temperature
50g caster sugar
1 large egg
100g self-raising flour
40ml skimmed milk at room temperature
20g vegetable oil
1/2 teaspoon pineapple essence
30g canned or fresh pineapple pieces

For Pineapple Buttercream

1 portion of Basic Buttercream
1/2 teaspoon pineapple essence
yellow food coloring
green food coloring

For Decoration

Open Star Piping Tip – for the main
  pineapple yellow part
Leaf Piping Tip – for pineapple stem part in
  green color
2 piping bags
pineapple paste for filling (instant can buy
  from bakery shop)

Cupcake Steps

1 Preheat the oven to 170 degrees Celsius.

2 Beat the butter and sugar until smooth.

3 Add in the egg and mix well.

4 Sift the self-raising flour.

5 Add in the flour little by little to the batter. Add in the milk little by little. Add in the vegetable oil and the pineapple essence. Stir in the crushed pineapple pieces.

6 Pour into cupcake cups, 2/3 full. Bake for 20-30 minutes or until golden brown.

7 Transfer to cooling rack. Cool down completely before frosting.

Pineapple Buttercream Icing Steps

1 Add the pineapple essence to 1 portion of Basic Buttercream.

2 Take 3/4 of buttercream, mix in yellow food coloring.

3 Take 1/4 remaining buttercream, mix in green food coloring.

Decoration Steps

1 Cut a hole in the middle of each cupcake. Fill centre with pineapple filling.

2 Place the Open Star Piping Tip into a piping bag. Place the Yellow Pineapple Buttercream into the piping bag and pipe onto each cupcake.

3 Place the Leaf Piping Tip into another piping bag. Place the Green Pineapple Buttercream into another piping bag and pipe the Green Pineapple Buttercream onto the top of the pineapple as the stem/leaves.

製作蛋糕

1 準備
將焗爐預熱至攝氏170度。牛油在室溫中回軟。

2 打發牛油
加入白砂糖，和牛油一起打發，直到顏色變淺，呈軟滑狀。

3 加入雞蛋
加打散的蛋液，混合攪拌，直至和牛油融合。

4 篩入粉類
將自發麵粉過篩，然後分3次加入。用切拌法快速拌勻。

5 加入其他材料
逐少加入脫脂牛奶。再分別加入植物油、鳳梨香油和罐頭或新鮮菠蘿粒。

6 入爐烘焙
慢慢倒入蛋糕杯，約2/3滿。用170度烘烤20-30分鐘或至蛋糕表面呈金黃色。

7 冷卻
完成後，轉移到散熱架上。等蛋糕完全冷卻後才裝飾。

製作鳳梨牛油糖霜

混合
將鳳梨香油加入一份預先準備的牛油糖霜中。

加入黃色色素
將3/4糖霜加入1-2滴黃色食用色素，拌勻。

加入綠色色素
將1/4糖霜加入1-2滴綠色食用色素，拌勻。

裝飾步驟

填餡
在蛋糕中間挖一個小孔，填入鳳梨餡。

使用開縫星齒唧嘴
先在蛋糕表面抹上一層黃色鳳梨牛油糖霜。再拿1個唧袋，放入開縫星齒唧嘴。將黃色鳳梨牛油糖霜放入唧袋，往下推實，紮緊袋口，在蛋糕面唧上星形花紋。

使用葉形唧嘴
再拿出1個唧袋，放入葉形唧嘴。將綠色鳳梨牛油糖霜放入唧袋，往下推實，紮緊袋口，在蛋糕面唧上葉形花紋。

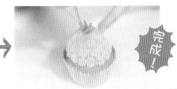

完成！

珍珠奶茶蛋糕
Bubble Tea Cupcakes

珍珠奶茶是我最喜歡的台灣特飲，
現在香港有很多地方都可以買到，不一定要飛去台灣。
不過我還是嫌不過癮，想嘗試新花樣，
於是就做了這個珍珠奶茶cupcake！
我用了黑茶、奶和黑糖香油，令蛋糕有珍珠奶茶的香味，
再加上黑珍珠，使口感更豐富。
這個蛋糕會裝飾成珍珠奶茶的樣子，
不過是一杯可以吃的珍珠奶茶哦！

小技巧 Tips

珍珠怎麼煮才更Q彈爽滑？

控制水量
將珍珠放入小鍋中，加入水量是珍珠的4倍，以中火煮到珍珠浮上面，改小火煮多10分鐘熄火。

熄火蓋上蓋子
蓋上蓋子焗10分鐘。

冷卻
將珍珠用飲用冷水沖瀝冷卻備用。

*在烘焙店可買到台灣真空脫水包裝的黑色珍珠粉圓。

製作蛋糕

1 準備
將焗爐預熱至攝氏170度。
牛油在室溫中回軟。

2 打發牛油
加入紅糖，和牛油一起打發，
直到顏色變淺，呈軟滑狀。

3 加入雞蛋
加入打散的蛋液，混合攪
拌，直至和牛油融合。

4 篩入粉類
將自發粉過篩，然後分3次
加入。用切拌法快速拌勻。

5 加入其他材料
逐少加入脫脂牛奶和黑茶。
加入植物油、黑糖香油和黑
珍珠粒。

6 入爐烘焙
慢慢倒入蛋糕杯，約2/3
滿。用170度烘烤20-30分
鐘或至表面呈金黃色。

7 冷卻
完成後，轉移到散熱架上。
等蛋糕完全冷卻後才裝飾。

裝飾步驟

使用開縫星齒唧嘴
拿1個唧袋，放入開縫星齒
唧嘴。將蛋白霜加入唧袋，
往下推實，紮緊袋口，在蛋
糕表面唧上花紋。

放上珍珠粒和飲管
放一些黑珍珠粒在蛋白霜上
面，再插一支飲管作裝飾。

材料 Ingredients

蛋糕
55克無鹽牛油(室溫)
50克紅糖
1隻雞蛋
100克自發粉
1茶包黑茶(用35毫升滾水泡焗
15分鐘)
10毫升脫脂牛奶
25克植物油
1/2茶匙黑糖香油
2湯匙(約30克黑珍
珠，烘焙店可買)

裝飾
開縫星齒唧嘴
1個唧袋
一份蛋白霜
珍珠粒
飲管作裝飾

For Cupcake
55g unsalted butter
50g light brown sugar
1 egg
100g self-raising flour
1 teabag black tea (soaked in 35ml hot
water for 15 mins)
10ml skimmed milk
25g vegetable oil
1/2 teaspoon black sugar essence
2 tablespoons black bubbles (30g) (buy
in bakery shop)

For Decoration
Open Star Piping Tip
1 piping bag
1 portion of Meringue Frosting
extra black bubbles
drinking straws

Cupcake Steps

1 Preheat oven 170 celcius.

2 Beat butter and sugar until smooth.

3 Add in beaten egg and mix well.

4 Sift the self-raising flour. Add in the
flour little by little to the batter.

5 Add in the milk and black tea little
by little. Add in the vegetable oil
and black sugar essence. Stir in
black bubbles.

6 Pour into cupcake cups, 2/3 full.
Bake for 20-30 minutes or until
golden brown.

7 Transfer to cooling rack. Cool
down completely before frosting.

Decoration Steps

1 Place the Open Star Piping Tip into
a piping bag. Place the Meringue
Frosting into the piping bag and
pipe onto the cupcake.

2 Topped with black bubbles.
Decorate with a drinking straw.

Tips: Preparation of the Black Bubbles

Steps: 1 Put dried tapioca balls into a pan, put in water (tapioca weight x 4 times). Bring
to boil until all tapioca balls starts floating on the surface of the water. Turn
down to medium heat and continue to cook for 10 minutes.

2 Turn the heat off and cover the pan with pan lid for another 10 minutes.

3 Wash the hot tapioca balls with drinking cold water (cold from fridge). Set aside
for later use.

*You can buy Tapioca Balls in baking supply stores called Taiwan Instant Dried Black
Tapioca Ball.*

黑糖薑母茶蛋糕

Ginger Cupcakes

薑母茶一般由老薑及黑糖或紅糖所製成。
以中醫的角度，薑母茶具有去風寒、開脾胃的功效，
所以很多人現在去台灣都會買這個作手信。
我當然不會錯過這麼受歡迎的口味！這款黑糖薑母茶cupcake，
表面選用另外一個我好喜歡吃的薑餅作裝飾，
相信是最有特色的自家製台灣手信！

材料 Ingredients

蛋糕

一份基本雲呢拿牛油蛋糕麵糊
1 湯匙薑粉（溶於 15 毫升熱水）
1/2 茶匙黑糖薑母茶粉
適量鹽

薑味牛油糖霜

100 克無鹽牛油（室溫）
180 克糖霜粉
20 克薑粉
2-3 茶匙脫脂牛奶

裝飾

開縫星齒唧嘴
1 個唧袋
薑餅曲奇

For Cupcake

1 portion of Basic Vanilla Cupcake
   Batter
1 tablespoon ginger powder (dissolved
   in 15ml hot water)
1/2 teaspoon brown sugar ginger tea
   powder
pinch of salt

For Ginger Buttercream

100g unsalted butter
180g icing sugar
20g ginger powder
2-3 teaspoon of skimmed milk

For Decoration

Open Star Piping Tip
1 piping bag
ginger cookie

製作蛋糕

1 混合麵糊和其他材料

準備一份基本雲呢拿牛油蛋糕麵糊，將薑粉和黑糖薑母茶粉溶液加入其中拌勻，再加少許鹽。

2 入爐烘焙

慢慢倒入蛋糕杯，約 2/3 滿，用 170 度烘烤 20-30 分鐘或至蛋糕表面呈金黃色。

3 冷卻

完成後，轉移到散熱架上，等蛋糕完全冷卻後才裝飾。

薑味牛油糖霜

打發牛油

將室溫牛油打發至軟綿。

加入糖霜、薑粉

分 5-6 次加入糖霜粉，加入薑粉。

加入牛奶

逐茶匙加入脫脂牛奶。

繼續打發

繼續打發至鬆軟順滑。

裝飾

使用開縫星齒唧嘴

拿 1 個唧袋，放入開縫星齒唧嘴。將薑味牛油糖霜加入唧袋，往下推實，紮緊袋口，從蛋糕表面唧上花紋。

放上薑餅曲奇

放上一塊薑餅曲奇作裝飾。

Cupcake Steps

1 Add in the ginger powder liquid and brown sugar ginger tea powder into Basic Vanilla Cupcake Batter. Add in a pinch of salt.

2 Pour into cupcake cups, 2/3 full. Bake for 20-30 minutes or until golden brown.

3 Transfer to cooling rack. Cool down completely before frosting.

Buttercream Steps

1 Beat the unsalted butter until soft.

2 Add in sugar and ginger powder and mix well.

3 Add in the milk, 1 teaspoon at a time.

4 Beat until smooth and creamy.

Decoration Steps

1 Place the Open Star Piping Tip into a piping bag. Place the Ginger Buttercream into the piping bag and pipe on to the cupcake.

2 Topped with ginger cookie.

紫薯蛋糕

Sweet Potato Cupcakes

這個紫番薯cupcake的糖霜非常健康，
因為用了紫番薯蓉製成。這個味道的靈感來自台灣九份，
當地把番薯叫作地瓜。在九份有很多地瓜做的食品，
例如地瓜酥、地瓜餅、地瓜芋圓等等，
我很想把這個味道帶回香港，就做了這個cupcake。

材料 Ingredients

蛋糕

一份基本雲呢拿牛油蛋糕
麵糊
1/2個紫薯（壓成泥醬，約
100克）
適量紫色的食用色素（可選）
10克甜紫薯乾（可選）

2-3茶匙脱脂牛奶
2個紫薯（壓成泥醬約
360克）

裝飾

多瓣花齒唧嘴
1個唧袋
紫薯乾（可選）
糖花
糖蝴蝶
紫色糖碎

For Cupcake

1 portion of Basic Vanilla
Cupcake Batter
1/2 small purple sweet
potatoes (mashed,
around 100g)
purple food coloring
(optional)
10g sprinkle dried sweet
potatoes (optional)

For Sweet Potato
Buttercream

100g unsalted butter
100g icing sugar
2-3 teaspoons of milk
2 small purple sweet potatoes
(mashed, around 360g)

For the Decoration

Closed Star Piping Tip
1 piping bag
dried sweet potato bits (optional)
suger flowers
suger butterflies
sprinkles

紫薯牛油糖霜

100克無鹽牛油（室溫）
100克糖霜粉

製作蛋糕

1 混合麵糊和紫薯
準備一份基本雲呢拿牛油蛋糕麵糊，將壓碎的紫薯加入其中。

2 加入其他材料
加入紫色的食用色素和紫薯乾。

3 入爐烘焙
慢慢倒入蛋糕杯，約2/3滿。用170度烘烤20-30分鐘或至蛋糕表面呈金黃色。

4 冷卻
完成後，轉移到散熱架上。等蛋糕完全冷卻後才裝飾。

製作甜薯泥醬

甜薯煲至軟熟
將甜薯放入小鍋中，加入熱水，以中火煲約30分鐘至完全淋軟熟透。

壓成泥醬
倒去水，待甜薯稍降溫除去外皮，用金屬湯匙背面將甜薯壓成泥醬備用。

*可用牙籤插入甜薯測試是否熟透。

製作紫薯牛油糖霜

打發牛油
將室溫牛油打至軟綿。

加入粉類
分 2-3 次加入糖霜粉。

加入其他材料
加入紫薯泥醬，拌匀；再逐茶匙加入脫脂牛奶。

繼續打發
打發至鬆軟滑身。

裝飾步驟

使用多瓣花齒唧嘴
拿1個唧袋，放入多瓣花齒唧嘴。將紫薯牛油糖霜加入唧袋，往下推實，紮緊袋口，在蛋糕表面唧上花紋。

放上糖花裝飾
放上糖飾花朵和蝴蝶，撒上紫色糖碎即可。

Cupcake Steps

1 Stir in the mashed purple sweet potatoes to 1 portion of Basic Vanilla Cupcake Batter.

2 Add a drop of purple food coloring if the batter is not purple enough. Stir in the dried sweet potatoes.

3 Pour into cupcake cups, 2/3 full. Bake for 20-30 minutes or until golden brown.

4 Transfer to cooling rack. Cool down completely before frosting.

Mashed Sweet Potato Paste Steps

1 Put one sweet potato into a pan and pour in hot water, enough to cover the entire sweet potato. Cook with medium heat for around 30 minutes until sweet potato becomes soften, checked with the insert of the toothpick or a small knife into the center.

2 Remove from water, let sweet potato cool down, then peel off the skin and mash it with a metal spoon.

Buttercream Steps

1 Beat the unsalted butter until soft.

2 Add in icing sugar and beat until smooth.

3 Add in the mashed sweet potatoes. Add in the milk,one teaspoon at a time.

4 Beat until smooth and creamy.

Decoration Steps

1 Place the Closed Star Piping Tip into a piping bag.

2 Place 1 portion of Sweet Potato Buttercream into a piping bag and pipe onto the cupcakes.

3 Decorate with sugar flowers, butterflies and sprinkles.

日本

JAPAN

日本人對食物的外觀和烹調方法很有要求。大家有沒有留意到日本人吃東西前，會將雙手合十，然後說「Itadakimasu」。這句話可以譯作「我不客氣了」。早在 1896 年日本著名養生學家石冢左玄已經在其著作《食物養生法》提出人類作為食物鏈的其中一環，對進食應抱持感恩態度。正因如此，他們在處理和烹調食物的時候都會很很用心。日本的食物包裝也非常精緻，我每每買回東西都不捨得拆開吃。這一章製作的 cupcake 除了吸收日本美食的經典口味外，也會在裝飾上體現細膩的和風！

好友
朱凱婷
的美食印象

一絲不苟是日本人的生活態度！從小到大我都很欣賞日本文化，走進日本的百貨公司，你會感受到各式各樣的產品是如何貼心地配合生活細節所需，由包裝到品質也是上盛之選！日本的飲食文化同樣追求「極上」享受！其中日本甜品及糕點「和菓子」是我的最愛，單是包裝已令人愛不釋手，材料方面更會因應季節而挑選。「期間限定」體現日本人精益求精的飲食追求：宇治綠茶、博多草莓、春天的櫻花等等全都能成為甜點的一部分！

朱凱婷

綠茶蛋糕

Green Tea Cupcakes

在日本，差不多每一間餐廳都會給你一杯綠茶，
特別是吃完甜品或是一些炸物之後，
喝一杯綠茶可以消滯。除此之外，
綠茶還有降膽固醇、消脂、防止蛀牙和
抗氧化等功效，確實有很多好處。
所以日本人每日都會喝綠茶，
而且很多甜品都會用綠茶做。

材料 Ingredients

蛋糕

45克無鹽牛油（室溫）
70克白砂糖
1隻雞蛋
90克自發粉
30毫升脫脂牛奶
15毫升綠茶（先溶於15毫升熱水內）
20克植物油
2茶匙綠茶粉

綠茶蛋白奶油霜

1份基本蛋白奶油霜
2湯匙綠茶粉

裝飾

1個開縫星齒唧嘴
1個唧袋
適量綠茶粉（撒面用）
綠茶朱古力手指餅

For Cupcake

45g unsalted butter, at room
  temperature
70g caster sugar
1 large egg
90g self-raising flour
30ml skimmed milk
15 ml green tea (green tea powder
  dissolved in 15ml hot water)
20g vegetable oil
2 teaspoon green tea powder

For Green Tea French Meringue Buttercream

1 portion of French Meringue
  Buttercream
2 tablespoons green tea powder

For Decoration

Open Star Piping Tip
1 piping bag
extra green tea powder to sprinkle on
  top
green tea chocolate coated biscuit
  sticks

Cupcake Steps

1 Preheat the oven to 170 degrees Celsius.

2 Beat the butter and sugar until smooth.

3 Add in the egg and mix well.

4 Sift the self-raising flour. Add in the flour little by little to the batter.

5 Add in the milk, green tea (water form), vegetable oil little by little. Stir in green tea powder.

6 Pour into cupcake cups, 2/3 full. Bake for 20-30 minutes or until golden brown.

7 Transfer to cooling rack. Cool down completely before frosting.

Green Tea French Meringue Buttercream Steps

1 Add the green tea powder into 1 portion of French meringue buttercream.

2 Beat until well blended and smooth.

Decoration Steps

1 Place the Open Star Piping Tip into a piping bag. Place the green tea powder into the Meringue Buttercream. Sprinkle with extra green tea powder on top of cupcake.

2 Decorate with 2 green tea chocolate coated biscuit sticks.

製作蛋糕

1 準備
將焗爐預熱至攝氏170度。牛油在室溫中回軟。

2 打發牛油
加入白砂糖，和牛油一起打發，直到顏色變淺，呈軟滑狀。

3 加入雞蛋
加打散的蛋液，混合攪拌，直至和牛油融合。

4 篩入粉類
將自發麵粉過篩，然後分3次加入。用切拌法快速拌勻。

5 加入其他材料
依次加入脫脂牛奶、綠茶溶液、植物油和綠茶粉。

6 入爐烘焙
慢慢倒入蛋糕杯，約2/3滿。用170度烘烤20-30分鐘或至蛋糕表面呈金黃色。

7 冷卻
完成後，轉移到散熱架上。等蛋糕完全冷卻後才裝飾。

製作綠茶蛋白奶油霜步驟

拌入綠茶粉
將綠茶粉均勻拌入預先打好的蛋白奶油霜內。

攪拌均勻
用電動攪拌器打發至滑身。

裝飾步驟

使用開縫星齒唧嘴
拿1個唧袋，放入開縫星齒唧嘴。將綠茶粉蛋白奶油霜加入唧袋，往下推實，紮緊袋口，在蛋糕表面唧上花紋。

用綠茶粉和綠茶朱古力手指餅裝飾
在奶油上撒上綠茶粉，再放上2枝綠茶朱古力手指餅作裝飾。

完成！

蜂蜜蛋糕
Honey Cupcakes

蜂蜜蛋糕通常都是長條形，切開幾片吃，
有濃厚的蜂蜜味道，質感也是軟綿綿的，
不過外形通常比較簡單。
所以我決定要做一個蜂蜜cupcake，彌補蜂蜜蛋糕
外形上的不足。蜂蜜當然是來自蜜蜂，
所以蛋糕表面用蜂蜜牛油霜唧成蜜蜂窩的樣子，
再加上小蜜蜂作裝飾，希望大家可以試試
這個與眾不同的蜂蜜蛋糕！

自家製蜜蜂裝飾

朱古力隔熱水座溶
分2個小碗，將無需調溫
白朱古力、無需調溫黑朱
古力隔熱水座溶。

畫出蜜蜂
將黑、白朱古力溶液分別放進兩個唧袋，在橢圓形黃色
糖果上面畫出蜜蜂形態。待朱古力變硬後才放上蛋糕面
作裝飾。

材料 Ingredients

蛋糕

55克無鹽牛油（室溫）
50克白砂糖
1隻雞蛋
100克自發粉
40毫升脫脂牛奶
25克植物油
20克蜂蜜

蜂蜜法式蛋白牛油糖霜

一份蛋白奶油霜
2湯匙蜂蜜
適量黃色食用色素

裝飾

圓口裱花唧嘴
1個唧袋
蜜蜂裝飾糖或橢圓形黃色糖果
無需調溫白朱古力
無需調溫黑朱古力

For the cupcake

55g unsalted butter, at room temperature
50g caster sugar
1 large egg
100g self-raising flour
40 ml skimmed milk
25g vegetable oil
20g honey

Honey French Meringue
Buttercream

1 portion of French Meringue Buttercream
2 teaspoons honey
yellow food coloring

Decoration

Round Piping Tip
1 piping bag
decorative bees or yellow oval shape
    candies
white chocolate (for hand drawn DIY
    decoration)
dark chocolate (for hand drawn DIY
    decoration)

製作蛋糕

1 準備
將焗爐預熱至攝氏170度。
牛油在室溫中回軟。

2 打發牛油
加白砂糖，和牛油一起打
發，至顏色變淺，呈軟滑狀。

3 加入雞蛋
加入打散的蛋液，混合攪
拌，直至和牛油融合。

4 篩入粉類
自發麵粉過篩，然後分3次
加入。用切拌法快速拌勻。

5 加入其他材料
逐小加入脫脂牛奶。加入植
物油。加入蜂蜜。

6 入爐烘焙
慢慢倒入蛋糕杯，約2/3
滿。用170度烘烤20-30分
鐘或至蛋糕表面呈金黃色。

7 冷卻
完成後，轉移到散熱架上。
等蛋糕完全冷卻後才裝飾。

Cupcake Steps

1 Preheat the oven to 170 degrees
  Celsius.
2 Beat the butter and sugar until
  smooth.
3 Add in the egg and mix well.
4 Sift the self-raising flour.
5 Add in the flour little by little to the
  batter. Add in the milk little by little.
  Add in the vegetable oil and the
  honey.
6 Pour into cupcake cups, 2/3 full.
  Bake for 20-30 minutes or until
  golden brown.
7 Transfer to cooling rack. Cool
  down completely before frosting.

製作蜂蜜法式蛋白奶油霜

拌入蜂蜜

將蜂蜜拌勻入預先準備好的
一份蛋白奶油霜內。

加入黃色色素

加入1-2滴黃色食用色素。
用手抹刀將顏色均勻地與奶
油霜混合。

裝飾步驟

使用圓口裱花唧嘴

拿1個唧袋，
放入圓口裱花
唧嘴。將蜂蜜
蛋白奶油霜放
入唧袋，往下
推實，紮緊袋
口，在蛋糕面唧上蜂巢花紋。

放上蜜蜂裝飾

放上蜜蜂裝飾糖（或自家製蜜
蜂裝飾）

Honey French Meringue
Buttercream Steps

1 Stir the Honey into the French
  Meringue Buttercream.
2 Add 1-2 drops of yellow food
  coloring. Mix until smooth and
  creamy.

Decoration Steps

1 Place the Round Piping Tip into
  a piping bag. Place the Honey
  French Meringue Buttercream into
  the piping bag and pipe onto each
  cupcake in a swirl motion.
2 Decorate with sugar bees on top
  of the buttercream.

Tips: Make your own D.I.Y. yellow candy bees with the use of Dark & white
Chocolate to hand drawn details.

Steps: 1 Put melted chocolate in a piping bag.
       2 Draw the bee outlines and details on top of oval shape candy.

Red Bean Cupcakes

大家小時候看卡通片裏面的卡通人物吃東西，
是不是想一起吃呢？我印象比較深刻的是多啦A夢的豆沙包。
其實日本人好喜歡用豆沙來做不同的和果子和甜品。
這個豆沙cupcake裏面除了豆沙外，還加入了豆沙味道的豆奶，
裝飾方面用了迷你版豆沙包，
讓大家重拾有多啦A夢相伴的童年回憶！

童年
的味道！

材料 Ingredients

蛋糕

55克無鹽牛油（室溫）
50克白砂糖
1隻雞蛋
100克自發粉
40毫升紅豆味豆奶（或低脂牛奶）
25克植物油
1/2茶匙香草香油
一罐紅豆餡

裝飾

多瓣花齒唧嘴
1個唧袋
200毫升淡忌廉
3個迷你豆沙餅（日本零食店可買到）
日本和服娃娃紙牌裝飾

For Cupcake

55g unsalted butter, at room temperature
50g caster sugar
1 large egg
100g self-raising flour
40ml red bean soy milk (or skimmed milk)
25g vegetable oil
1/4 teaspoon vanilla essence
1 can red bean paste

For Decoration

Closed Star Piping Tip
1 piping bag
200 ml whipped cream
3 mini red bean buns (can buy in japan snacks shop)
Japanese doll paper toppers

製作蛋糕

1 準備
將焗爐預熱至攝氏170度。牛油在室溫中回軟。

2 打發牛油
加入白砂糖，和牛油一起打發，至顏色變淺，呈軟滑狀。

3 加入雞蛋
加入打散的蛋液，混合攪拌，直至和牛油融合。

4 篩入粉類
自發麵粉過篩，然後分3次加入。用切拌法快速拌勻。

5 加入其他材料
逐少加入紅豆味豆奶、植物油和香草香油。

6 入爐烘焙
將蛋糕麵糊慢慢倒入蛋糕杯，約2/3滿。用170度烘烤20-30分鐘或至蛋糕表面呈金黃色。

7 冷卻
完成後，轉移到散熱架上。等蛋糕完全冷卻後才裝飾。

8 填餡
在每一個蛋糕中間開一個小孔。將紅豆餡填入小孔。

裝飾步驟

打發淡忌廉
將淡忌廉打企身。

使用多瓣花齒唧嘴
拿1個唧袋，放入多瓣花齒唧嘴。將鮮忌廉加入唧袋，往下推實，紮緊袋口，在蛋糕表面唧上花紋。

加上紅豆餡和豆沙夾餅
放一點點紅豆餡在鮮忌廉的中間。放半個豆沙夾餅在鮮忌廉上面。

插上日本和服娃娃紙牌
最後，插一個日本和服娃娃紙牌裝飾。

Cupcake Steps

1 Preheat the oven to 170 degrees Celsius.

2 Beat the butter and sugar until smooth.

3 Add in the egg and mix well.

4 Sift the self-raising flour. Add in the flour little by little to the batter.

5 Add in the red bean soya milk little by little. Add in the vegetable oil and vanilla essence.

6 Pour into cupcake cups, 2/3 full. Bake for 20-30 minutes or until golden brown.

7 Transfer to cooling rack. Cool down completely before frosting.

8 Cut a hole in the middle of the cupcakes. Fill with red bean paste.

Decoration Steps

1 Whip the cream to peak.

2 Place the Closed Star Piping Tip into a piping bag. Place the Whipped Cream into the piping bag and pipe onto each cupcake.

3 Place small amounts of red bean paste in centre of each cupcake.

4 Decorate with half piece of red bean bun. Decorate with Japanese doll paper topper.

巨峰提子蛋糕

Grape Cupcakes

提子之中，我最喜歡巨峰提子，因為它有獨特的味道。
秋天直至11月是巨峰提子的收成期，
當中便以有水果王國美譽的山梨縣為最佳。若你鍾情巨峰，
一久園可以容許你在園內任吃四十分鐘，費用也只需大約130港元。
但如果去不到日本，在香港也有很多地方可以買到。
這個cupcake加入了我最喜歡的巨峰提子粒，
外形設計上也別具新意，大家不妨動手做做看！

材料 Ingredients

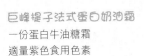

蛋糕
45克無鹽牛油（室溫）
70克白砂糖
1隻雞蛋
90克自發粉
30毫升脫脂牛奶
20克植物油
10毫升鮮巨峰提子汁
20克鮮巨峰提子粒

巨峰提子法式蛋白奶油霜
一份蛋白牛油糖霜
適量紫色食用色素
1/2茶匙提子香油

裝飾
提子軟糖
綠色葉形軟糖（或提子綠色莖部份）

For Cupcake
45g unsalted butter, at room temperature
70g caster sugar
1 large egg
90g self-raising flour
30ml skimmed milk
20g vegetable oil
10ml fresh grape juice
20g chopped grape pieces

For Grape French Meringue Buttercream
1 portion of French Meringue Buttercream
1 drop of purple food coloring
1/2 teaspoon grape essence

For Decoration
Round Piping Tip
round or oval shape grape sugar coated soft candies
oval shape green sugar coated soft candies (or stem of the grapes)

製作蛋糕

1 準備
將焗爐預熱至攝氏170度。牛油在室溫中回軟。

2 打發牛油
加入白砂糖，和牛油一起打發，至顏色變淺，呈軟滑狀。

3 加入雞蛋
加入打散的蛋液，混合攪拌，直至和牛油融合。

4 篩入粉類
自發麵粉過篩，然後分3次加入。用切拌法快速拌勻。

5 加入其他材料
逐少加入脫脂牛奶。加入植物油。加入鮮巨峰提子汁。加入鮮巨峰提子粒。

6 入爐烘焙
慢慢倒入蛋糕杯，約2/3滿。用170度烘烤20-30分鐘或至蛋糕表面呈金黃色。

7 冷卻
完成後，轉移到散熱架上。等蛋糕完全冷卻後才裝飾。

Cupcake Steps
1 Preheat the oven to 170 degrees Celcius.
2 Beat the butter and sugar until smooth.
3 Add in the egg and mix well.
4 Sift the self-raising flour. Add in the flour little by little into the batter.
5 Add in the milk little by little. Add the vegetable oil. Add in fresh grape juice. Add in grapes.
6 Pour into cupcake cups, 2/3 full. Bake for 20-30 minutes or until golden brown.
7 Transfer to cooling rack. Cool down completely before frosting.

製作巨峰提子法式蛋白奶油霜

加入色素和提子香油
將紫色食用色素和提子香油拌勻，加入預先打好的蛋白奶油霜。

打發奶油霜
用電動攪拌器打發至順滑。

裝飾

抹平奶油霜
用手抹刀或小匙將奶油霜抹平在蛋糕面上。

放上綠色莖
拿一塊綠色軟糖，或剪下提子綠色莖部份，放上蛋糕面，裝飾成提子莖部。

放上提子糖果
放提子糖果在奶油霜上面，排列成一束提子的樣子。

Grape French Meringue Buttercream Steps
1 Add in the grape essence and purple food coloring into 1 portion of French Meringue Buttercream.
2 Beat until well blended.

Decoration Steps
1 Spread the Grape French Meringue Buttercream onto each cupcake using a spatula or a spoon.
2 Place oval shape green sugar coated soft candies on each cupcake or cut and trim the stems of the grapes into short mini stems and attached to each cupcake.
3 Place the grape candies on top of each cupcake and arrange like a bunch of grapes.

新加坡·泰國
SINGAPORE THAILAND

新加坡和泰國給人一種度假的感覺：陽光與海灘，無拘無束，很悠閒。出街的時候不用刻意打扮，T恤、短褲加一對拖鞋就可以。熱帶的天氣和沿海的地理位置，造就了當地新鮮的食材，許多菜式都會加入當地常見的植物：比如泰國的雞飯和蕉葉，新加坡的海南雞飯和班蘭葉。兩地新鮮的水果、香草、海鮮也是遊客必嚐的美食！

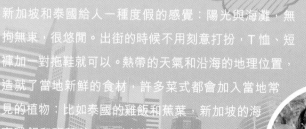

好友
朱凱婷
的美食印象

新加坡同泰國是亞熱帶地方，全年都是夏季。熱帶地區有一樣好處就是有很多新鮮熱帶水果，所以當地的食物會用很多水果來做。由於天氣炎熱，當地人除了喜歡吃海鮮之外，也喜歡吃甜品來消暑。我最喜歡的食品就是班蘭葉做的班蘭蛋糕。這個是我每次去新加坡都一定會吃的。

朱凱婷

芒果蛋糕
Mango Cupcakes

常常都會見到芒果味蛋糕，
而且一定是用新鮮的芒果填滿蛋糕的頂部。
除了蛋糕外，還有芒果撻、芒果卷、芒果批，
但是好像很少見到芒果味的 cupcake，
所以我做了這個 cupcake。
我選用新鮮芒果肉砌成一朵芒果花，
用來送給女孩子最合適不過！
不過記得要選擇一些比較甜的芒果哦！

材料 Ingredients

蛋糕

55克無鹽牛油（室溫）
50克白砂糖
1隻雞蛋
100克自發粉
30毫升鮮芒果汁
10毫升脫脂牛奶
25克植物油
1/2茶匙芒果精油
15克芒果乾（可選）

裝飾

100毫升淡忌廉
2-3個新鮮芒果切片

For Cupcake

55g unsalted butter, at room temperature
50g caster sugar
1 large egg
100g self-raising flour
30ml fresh mango juice
10ml skimmed milk
25g vegetable oil
1/2 teaspoon mango essence
15g dried mango bits (optional)

For Decoration

100ml whipped cream
2-3 fresh mangos (cut into slices)

Cupcake Steps

1 Preheat the oven to 170 degrees Celsius.

2 Beat the butter and sugar until smooth.

3 Add in the egg and mix well.

4 Sift the self-raising flour. Add in the flour little by little to the batter.

5 Add in the fresh mango juice and milk little by little. Add in the vegetable oil. Add in the mango essence and dried mangoes

6 Pour into cupcake cups, 2/3 full. Bake for 20-30 minutes or until golden brown.

7 Transfer to cooling rack. Cool down completely before frosting.

Decoration Steps

1 Whip the cream to peak.

2 Frost the top of the cupcake with thin layer of whipped cream.

3 Slice the mangos into small pieces.

4 Arrange mango slices around cupcake in shape of flower petals.

Tips: Use Sweeter mangos.

 **小技巧 Tips** 使用甜芒果，味道更好！

製作蛋糕

1 準備
將焗爐預熱至攝氏170度。牛油在室溫中回軟。

2 打發牛油
加入白砂糖，和牛油一起打發，直到顏色變淺，呈軟滑狀。

3 加入雞蛋
加入打散的蛋液，混合攪拌，直至和牛油融合。

4 篩入粉類
將自發麵粉過篩，然後分3次加入。用切拌法快速拌勻。

5 加入其他材料
逐少加入芒果汁和牛奶。加入的植物油。加入芒果精油和芒果乾碎。

6 入爐烘焙
慢慢倒入蛋糕杯，約2/3滿。用170度烘烤20-30分鐘或至蛋糕表面呈金黃色。

7 冷卻
完成後，轉移到散熱架上。等蛋糕完全冷卻後才裝飾。

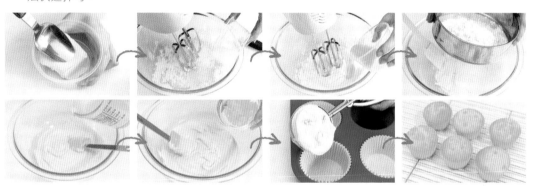

裝飾

打發鮮忌廉
將鮮忌廉打企身。

塗抹蛋糕
將鮮忌廉抹塗在蛋糕上面。

芒果切片
芒果切成花瓣形小片，用紙巾輕壓，吸乾水分。

擺面
將芒果切片沿蛋糕中心開始向外圍繞，砌成花形。

椰子蛋糕
Coconut Cupcakes

一提到椰子，就會想到海島上的陽光與海灘。
作為熱帶島嶼的特產，椰子不僅美味，而做
法多樣，可以製成椰子油、椰汁、椰子
糖……做這個味道的cupcake，我用
了椰子糖、椰子肉、椰奶和椰
絲，來確保cupcake充滿濃烈
的椰子味。

材料 Ingredients

蛋糕
55克無鹽牛油（室溫）
50克白砂糖
1隻雞蛋
100克自發粉
40毫升椰奶
25克植物油
20克椰絲碎

椰子牛油糖霜
100克無鹽牛油（室溫）
200克糖霜粉
2-3茶匙椰奶
10克椰絲碎

裝飾
多瓣花齒唧嘴
1個唧袋
烤焗椰絲
椰子糖漿
6個裝飾小紙傘
3粒麥提莎朱古力（分為
　兩半）

For Cupcake
55g unsalted butter, at room
　temperature
50g caster sugar
1 large egg
100g self-raising flour
40ml coconut milk
25g vegetable oil
20g desiccated coconut or
　coconut shreds

For Coconut Buttercream
100g unsalted butter, at room
　temperature
200g icing sugar
2-3 teaspoons coconut milk
10g desiccated coconut

For Decoration
Closed Star Piping Tip
1 piping bag
toasted coconut
coconut syrup
6 small party paper umbrella
3 pieces of Maltesers
　chocolate (cut into halves)

製作蛋糕

1 準備
將焗爐預熱至攝氏170度。牛油在室溫中
回軟。

2 打發牛油
加入白砂糖，和牛油一起打發，直到顏色變
淺，呈軟滑狀。

3 加入雞蛋
加打散的蛋液，混合攪拌，直至和牛油融合。

4 篩入粉類
將自發麵粉過篩，然後分3次加入。用切拌
法快速拌勻。

5 加入其他材料
逐少加入椰奶。
加入植物油和
椰絲碎。

6 入爐烘焙
慢慢倒入蛋糕
杯，約2/3滿。

在蛋糕麵糊中加入椰絲碎

用170度烘烤20-30分鐘或至蛋糕表面呈金
黃色。

7 冷卻
完成後，轉移到散熱架上。等蛋糕完全冷卻
後才裝飾。

製作椰子牛油糖霜

打發牛油
將室溫牛油打至軟綿，同時
分6-7次加入糖粉。 →

加入椰奶和椰絲
逐茶匙加入椰奶。加入椰絲
碎拌勻。 →

繼續打發
用電動攪拌器打發至鬆軟順
滑。

裝飾

使用多瓣花齒唧嘴
拿1個唧袋，放入多瓣花齒唧
嘴。將椰子牛油糖霜加入唧袋，
往下推實，紮緊袋口，在蛋糕表
面唧上花紋。 →

加上椰絲和
椰子糖漿
撒上烤焗椰
絲，淋上一
些椰子糖漿。 →

放上朱古力裝飾
將半粒麥提莎朱古力
裝飾成椰子模樣，放
在蛋糕面上。 →

插上小紙傘
插一把小紙
傘作裝飾。

Cupcake Steps

1 Preheat the oven to 170 degrees
Celsius.

2 Beat the butter and sugar until
smooth.

3 Add in the egg and mix well.

4 Sift the self-raising flour. Add in the
flour little by little to the batter.

5 Add in the coconut milk little by
little. Add in the vegetable oil.
Stir in the desiccated coconut or
coconut shreds.

6 Pour into cupcake cups, 2/3 full.

Bake for 20-30 minutes or until
golden brown.

7 Transfer to cooling rack. Cool
down completely before frosting.

Coconut Buttercream Steps

1 Beat the unsalted butter until soft.
Sift the icing sugar and add in 6-7
times.

2 Add in the coconut milk, one
teaspoon at a time. Stir in the
desiccated coconut.

3 Beat until smooth and creamy.

Decoration Steps

1 Place the Closed Star Piping
Tip into a piping bag. Place the
Coconut Buttercream into the
piping bag and pipe onto each
cupcake.

2 Sprinkled with Toasted Coconut.
Drizzle some Coconut Syrup on
each cupcake.

3 Decorate with 1/2 a Maltesers
chocolate.

4 Decorate with a party paper
umbrella.

咖椰蛋糕
Kaya Cupcakes

咖椰多士是我去新加坡必食的美味！
我覺得咖椰介於花生醬同蜜糖之間，
後兩者常常被用於製作甜點，
所以我想咖椰或許是做cupcake的好材料！
大家可以試試這個食譜，
這樣以後除了用咖椰做多士之外，
還有另一個選擇！

材料 Ingredients

蛋糕
55克無鹽牛油（室溫）
50克白砂糖
1隻雞蛋
100克自發粉
40毫升椰奶
25克植物油
1/2湯匙咖椰醬

咖椰牛油糖霜
100克無鹽牛油（室溫）
180克糖霜粉

2-3茶匙椰奶
2茶匙咖椰醬

裝飾
開縫法式裱花唧嘴
1個唧袋
6塊迷你多士
咖椰醬（搽上多士面）

For Cupcake
55g unsalted butter, at room
   temperature
50g caster sugar
1 large egg
100g self-raising flour, sifted
40ml coconut milk
25g vegetable oil
2 tablespoons kaya paste

For Kaya Buttercream
100g unsalted butter, at room
   temperature
180g icing sugar
2-3 teaspoons coconut milk
2 teaspoons kaya paste

For Decoration
French Star Piping Tip
1 piping bag
6 pieces of mini toasted bread
extra kaya paste for cake filling
   and spreading on toasted
   bread

製作蛋糕

1 準備
將焗爐預熱至攝氏170度。牛油在室溫中回軟。

2 打發牛油
加白砂糖，和牛油一起打發，至顏色變淺，呈軟滑狀。

3 加入雞蛋
加打散的蛋液，混合攪拌，直至和牛油融合。

4 篩入粉類
自發麵粉過篩，然後分3次加入。用切拌法快速拌勻。

5 加入其他材料
逐少加入椰奶、咖椰醬和植物油。

6 入爐烘焙
慢慢倒入蛋糕杯，約2/3滿。用170度烘烤20-30分鐘或至蛋糕表面呈金黃色。

7 冷卻
完成後，轉移到散熱架上。等蛋糕完全冷卻後才裝飾。

製作咖椰牛油糖霜

打發牛油
將室溫牛油打至軟綿，分6-7次加入糖霜粉。 →

加入椰奶
逐茶匙加入椰奶。 →

加入咖椰醬
加入咖椰醬攪勻。 →

繼續打發
打發至鬆軟滑身。

裝飾

填入咖椰醬
在蛋糕中間挖一小孔，填入咖椰醬。 →

使用開縫星齒唧嘴
拿1個唧袋，放入開縫法式裱花唧嘴。將咖椰牛油糖霜放入唧袋，往下推實，紮緊袋口，在蛋糕表面唧上花紋。 →

放上咖椰多士
將咖椰多士切小塊，每一塊小多士搽上一層薄咖椰醬。將咖椰多士放在每一個蛋糕上作裝飾。

Cupcake Steps

1 Preheat the oven to 170 degrees Celsius.

2 Beat the butter and sugar until smooth.

3 Add in the egg and mix well.

4 Sift the self-raising flour. Add in the flour little by little to the batter.

5 Add in the coconut milk little by little. Add in the vegetable oil. Stir in the kaya paste.

6 Pour into cupcake cups, 2/3 full. Bake for 20-30 minutes or until golden brown.

7 Transfer to cooling rack. Cool down completely before frosting.

Kaya Buttercream Steps

1 Beat the unsalted butter until soft. Sift the icing sugar and add in 6-7 times.

2 Add in the milk one teaspoon at a time.

3 Stir in the Kaya paste.

4 Beat until smooth and creamy.

Decoration Steps

1 Cut a hole in each cupcake. Fill each centre with kaya paste.

2 Place the French Star Piping Tip into a piping bag. Place the Kaya Buttercream into the piping bag and pipe onto each cupcake.

3 Spread some kaya paste on each mini toast. Decorate with Mini Kaya Toast on each cupcake.

班蘭蛋糕

Pandan Cupcakes

班蘭蛋糕是新加坡的特產，也是最受歡迎的手信！
在機場已經有很多店舖可以買到。班蘭葉有一種獨特的
天然香氣，能給食物增添清新香甜的味道。
這種植物在新加坡非常普遍，
當地人會用在雞飯、果醬、麵包、曲奇裏，
確實有很多用途！現在我把班蘭味道放進cupcake，
這樣在家也可隨時吃到星洲美味啦！

濃濃的
班蘭香氣～

材料 Ingredients

蛋糕
45克無鹽牛油(室溫)
70克白砂糖
1隻雞蛋
90克自發粉
40毫升椰奶
20克植物油
1/2茶匙班蘭香油
適量綠色食用色素

班蘭法式蛋白奶油霜
1份法式蛋白奶油霜
1/2茶匙班蘭香油
1-2滴綠色食用色素

裝飾
開縫星齒唧嘴
1個唧袋
6支小飲管
適量糖花

For Cupcake
45g unsalted butter, at room temperature
70g caster sugar
1 large egg
90g self-raising flour
40ml coconut milk at room temperature
20g vegetable oil
1/2 teaspoon pandan essence
green food coloring (optional)

For Pandan French Meringue Buttercream
1 portion of French Meringue
  Buttercream
1/2 teaspoon pandan essence
1drop of green food coloring

For Decoration
Open Star Piping Tip
1 piping bag
6 drinking straws
sugar flowers

製作蛋糕

1 準備
將焗爐預熱至攝氏170度。牛油置於室溫中回軟。

2 打發牛油
加白砂糖，和牛油一起打發，至顏色變淺，呈軟滑狀。

3 加入雞蛋
加入打散的蛋液，混合攪拌，直至和牛油融合。

4 篩入粉類
自發麵粉過篩，然後分3次加入。用切拌法快速拌勻。

5 加入其他材料
逐少加入椰奶。再加入植物油和班蘭香油。

6 入爐烘焙
慢慢倒入蛋糕杯，大約2/3滿。用170度烘烤20-30分鐘或至蛋糕表面呈金黃色。

7 冷卻
完成後，將蛋糕轉移到散熱架上。等蛋糕完全冷卻後才裝飾。

製作班蘭法式蛋白奶油霜

加入班蘭香油
將班蘭香油加入預先備好的法式蛋白奶油霜中。

加入色素
加1-2滴綠色食用色素。用手抹刀將顏色與奶油霜混和均勻。

裝飾

使用開縫星齒唧嘴
拿1個唧袋，放入開縫星齒唧嘴。將班蘭法式蛋白奶油霜放入唧袋，往下推實，紮緊袋口，在蛋糕表面唧上花紋。

放上裝飾物
在每一個蛋糕上放上一支小吸管和幾朵糖花作裝飾。

Cupcake Steps
1 Preheat the oven to 170 degrees Celsius.
2 Beat the butter and sugar until smooth.
3 Add in the egg and mix well.
4 Sift the self-raising flour. Add in the flour little by little to the batter.
5 Add in the coconut milk little by little. Add in the vegetable oil and the pandan essence.
6 Pour into cupcake cups, 2/3 full. Bake for 20-30 minutes or until golden brown.
7 Transfer to cooling rack. Cool down completely before frosting.

Pandan French Meringue Buttercream Steps
1 Add the pandan essence into 1 portion of French Meringue Buttercream.
2 Add in the green food coloring. Mix until smooth and creamy.

Decoration Steps
1 Place the Open Star Piping Tip into a piping bag. Place the Pandan Buttercream into the piping bag and pipe onto each cupcake.
2 Decorate with drinking straw and sugar flower.

英國
UNITED
KINGDOM

說起英國飲食，相信大家最熟悉的就是英式早餐和 high tea。不過很多人誤以為 high tea 是下午茶的意思，其實傳統英國人所說的"high tea"或"meat tea"是晚餐。相反，下午茶應該叫做"low tea"，而且還分為三類。第一類"cream tea"只有茶、鬆餅和果醬；第二類"light tea"會再加少許甜點；而第三類"full tea"通常會加三文治、蛋糕、曲奇餅等等。英國菜烹調方法通常比較簡單，但這也許正是他們的特色。對於餐桌禮儀，英國人則相當講究，運用的餐具都大有名堂，這也是英國成為優雅、高貴的代名詞的原因吧！

好友
陳茵媺
的美食印象

傳統英國的皇族貴氣，給人留下最深刻印象。傳統英式烤餅配以英式香茶，體現英國人對生活品味和格調的追求。格雷伯爵茶，黃金古瓷杯，明火暖鮮奶，hmm……這感覺要去形容是很難的，而要忘記它，更難！

陳茵媺

UNITED KINGDOM
英國

檸檬批蛋糕

Lemon Meringue Pie Cupcakes

檸檬批是我媽媽的最愛。記憶中這個甜品有很多忌廉和很香
的檸檬味。為了保留這份甜蜜回憶,我將它做成杯子蛋糕。
很多人爭拗檸檬批的起源,有人說源自英國,有人說是法
國,也有人說是美國,但還沒有定論。不過,網上有一個
比較接近的說法,就是早在1940年,英國已經有一個叫做
"Chester pudding" 的甜品跟檸檬批很相似。

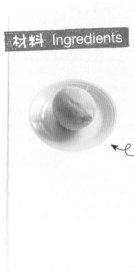

材料 Ingredients

脆餅底
40克無鹽牛油（溶化）
80克消化餅乾（壓碎）

蛋糕
一份基本雲呢拿牛油蛋糕麵糊
2湯匙檸檬汁
1/2個檸檬（皮）

檸檬酪醬
1隻雞蛋
20毫升檸檬汁
1/4湯匙檸檬皮
30克白砂糖
15克無鹽牛油

裝飾
開縫星齒唧嘴
1個唧袋
一份蛋白霜
鮮檸檬切片或檸檬形狀糖果

For Bottom Layer
40g unsalted butter (melted)
80g digestive biscuits (crumbs)

For Cupcake
1 portion of Basic Vanilla Cupcake Batter
2 tablespoon lemon juice
zest of 1/2 lemon

For Lemon Filling
1 large egg
20ml freshly squeezed lemon juice
1/4 tablespoon finely shredded lemon zest
30g granulated white sugar
15g room temperature unsalted butter

For Decoration
Open Star Piping Tip
1 piping bag
1 portion of Meringue Frosting
Fresh lemon slices or lemon jelly candies

Cupcake Steps

1. For the Bottom: Melt the butter, add in digestive biscuits to make into crumbs.
2. Place the mixture and press into flat layer 4-5mm on bottom of each cupcake.
3. For the Cupcake: Add the lemon juice and lemon zest to 1 portion of Basic Vanilla Cupcake Batter.
4. Pour into cupcake cups, 2/3 full. Bake for 30-40 minutes or until golden brown.
5. Transfer to cooling rack. Cool down completely before frosting.

Filling Steps

1. Whisk the eggs, sugar and lemon juice in a mixing bowl over a saucepan of simmering water, until blended. Stir constantly until mixture becomes thickened (around 10 minutes).
2. Remove from heat, pour through fine strainer to remove lumps.
3. Cut the butter in small pieces and whisk it into the mixture until co-operated.
4. Add in the lemon zest.
5. Let it cool and refrigerate overnight.
6. Cut a hole in each cupcake. Fill each centre with lemon filling.

Decoration Steps

1. Place the Open Star Piping Tip into a piping bag. Place the Meringue Frosting into the piping bag and pipe onto each cupcake.
2. Use a flaming torch to slightly burn meringue to light brown color.
3. Decorate with a slice of fresh lemon or lemon jelly candy.

製作蛋糕

1 製作脆餅底
將牛油溶解,加入餅乾碎。將餅乾碎壓平4-5mm在蛋糕杯的底部。

2 製作檸檬麵糊
將檸檬汁和檸檬皮拌勻,倒一份基本雲呢拿牛油蛋糕麵糊裏。

3 入爐烘焙
慢慢倒入蛋糕杯,約2/3滿。用170度烘烤20-30分鐘或至蛋糕表面呈金黃色。

4 冷卻
完成後,轉移到散熱架上。等蛋糕完全冷卻後才裝飾。

製作檸檬酪醬

隔熱水煮溶
將雞蛋、糖、檸檬汁放在碗中拌勻,然後在鍋裏隔熱水煮溶。離火,不斷攪拌10分鐘,直到混合物變得濃稠。

過濾
將混合物過濾,以去除團塊。

加入牛油
將牛油切小塊加入混合物中,攪拌至完全融合。

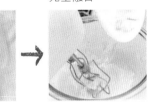

加入檸檬皮
將檸檬皮倒入混合物中,攪拌均勻。

入雪櫃冷藏
待它冷卻後,放入雪櫃冷藏一天備用。

填餡
在蛋糕中間挖一小孔,填入檸檬酪醬。

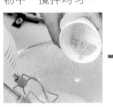

*也可到超市買即食檸檬蛋黃醬。

裝飾

使用開縫星齒唧嘴
拿1個唧袋,放入開縫星齒唧嘴。將蛋白霜放入唧袋,往下推實,紮緊袋口,在蛋糕表面唧上花紋。

將蛋白霜表面燒至金黃色
使用烘焙火槍,將蛋白霜表面燒至金黃色。

放上檸檬
最後,放一塊鮮檸檬片當裝飾。

完成

麵包布甸蛋糕
Bread Pudding Cupcakes

麵包布甸是很多傳統英國家庭都喜歡的甜品！
原因是簡單易做，只需要用剩餘的麵包混合其他材料，
不僅操作容易，還可以避免食物的浪費。
我自己喜歡用提子麵包，加入少少雞蛋、忌廉、
牛奶和肉桂粉去做麵包布甸。
於是，我將這個簡單的麵包布甸食譜融入cupcake，
製造一個充滿英式風味的麵包布甸蛋糕。

材料 Ingredients

蛋糕
一份基本雲呢拿牛油蛋糕麵糊
1/4 杯提子乾

麵包布甸
3 塊提子乾麵包
1 茶匙肉桂粉
1 隻雞蛋
40 毫升牛奶
10 毫升淡忌廉

For Cupcake
1 portion of Basic Vanilla Cupcake Batter
1/4 cup raisins

For Bread Pudding Topping
3 pieces of raisin bread
1 teaspoon grounded cinnamon powder
1 egg
40ml milk
10ml whipping cream

製作步驟

1 放入提子乾
將提子乾放入一份預先備好的基本雲呢拿牛油蛋糕麵糊裏。將麵糊慢慢倒入蛋糕杯，約2/3滿。

2 麵包切片
將提子乾麵包切成小方片。

3 混合
將肉桂粉、雞蛋、牛奶和鮮忌廉倒入一個大碗內，混合均勻。然後將提子麵包片放入其中。

4 放在蛋糕頂部
將浸透的小方片麵包放在蛋糕的頂部。

5 入爐烘焙
在蛋糕麵糊上平鋪麵包布丁，入爐用170度烘烤20-30分鐘或至蛋糕表面呈金黃色。

Cupcake Steps

1 Stir in the raisins to 1 portion of Basic Vanilla Cupcake Batter. Pour into cupcake cups, 2/3 full.
2 Cut the bread into small square pieces.
3 Mix the grounded cinnamon, egg, milk and cream in a cup. Dip the bread pieces into the cream mixture.
4 Place a few pieces on top of the cupcakes.
5 Bake together with the cupcakes for 20-30 minutes or until golden brown.

格雷伯爵茶蛋糕

Earl Grey Tea Cupcakes

傳說格雷伯爵茶的起源是因為格雷伯爵
依自己的喜好將不同的茶葉混合，並加入具安定效用的
香檸檬油，之後傳入民間，廣受大眾歡迎。
而英國人很喜歡嘆下午茶，他們常常會一邊喝茶，一邊吃甜點。
所以我從中得到靈感，將格雷伯爵茶加入蛋糕裏面，
再將蛋糕放進茶杯裏面，做出一杯食得的「茶」，
一次過滿足兩個要求！

材料 Ingredients

蛋糕

一份基本雲呢拿牛油蛋糕麵糊
20毫升格雷伯爵茶（用
　熱水泡焗）
1茶匙格雷伯爵茶葉

裝飾

開縫星齒唧嘴
1個唧袋
250毫升淡忌廉
格雷伯爵茶葉（撒面用）

For the cupcake

1 portion of Basic Vanilla Cupcake
　Batter
20ml Earl Grey tea (soaked in hot
　water)
1 teaspoon Earl Grey tea leaves

For the Decoration

Open Star Piping Tip
1 piping bag
250ml whipped cream
Earl Grey tea leaves (sprinkle on top)

製作蛋糕

1 加入格雷伯爵茶
將格雷伯爵茶用熱水泡焗，
泡好後逐少倒入預先備好
的基本雲呢拿牛油蛋糕麵
糊中。

2 加入茶葉
將格雷伯爵茶葉倒入蛋糕
糊，拌勻。

3 入爐烘焙
將麵糊慢慢倒入蛋糕杯，
約2/3滿。用170度烘烤
20-30分鐘或至蛋糕表面呈
金黃色。

4 冷卻
完成後，轉移到散熱架上。
等蛋糕完全冷卻後才裝飾。

裝飾步驟

打發忌廉
將淡忌廉打發企身。

撒上格雷伯爵茶葉
在蛋糕面上再撒一些格雷伯
爵茶葉作裝飾。

使用開縫星齒唧嘴
拿1個唧袋，放入開縫星齒
唧嘴。將鮮忌廉抹入唧袋，
往下推實，紮緊袋口，在蛋
糕表面唧上花紋。

放入茶杯模具
最後，把蛋糕放入特別的茶
杯模具內，美化外觀。

Cupcake Steps

1 Add in the Earl Grey tea (water)
　little by little into 1 portion of Basic
　Vanilla Cupcake Batter.

2 Stir in the Earl Grey tea leaves
　(dried).

3 Pour into cupcake cups, 2/3 full.
　Bake for 20-30 minutes or until
　golden brown.

4 Transfer to cooling rack.Cool down
　completely before frosting.

Decoration Steps

1 Whip the cream until peak.

2 Place the Open Star Piping Tip into
　a piping bag. Place the whipped
　cream into the piping bag and pipe
　onto each cupcake.

3 Sprinkled with dried Earl Grey tea
　leaves.

4 Served on teacup.

香橙朱古力蛋糕

Jaffa Cake Cupcakes

香橙加朱古力可以說是最佳配搭，而英國人最愛的甜食
Jaffa cakes就是香橙朱古力加蛋糕。為了做一個cupcake版本，
我選擇用朱古力蛋糕做底，加入新鮮的橙皮、香橙果醬和
橙汁來製造濃烈的香橙味道，而頂部則用了朱古力醬！
最後，將新鮮香橙塊焗至乾身，再沾上朱古力。
這樣就可以確保香橙味和朱古力味一樣濃烈！

**小技巧
Tips**

如何製作香橙朱古力片？

切橙片
將橙切成大約3mm
薄片，然後把橙片
平鋪在烘焙紙上。

焗乾
放入焗爐低溫度
80℃焗1.5至2小
時，直到水分乾涸。

朱古力隔熱水座溶
將黑朱古力放一個
碗裏，隔熱水座溶。

沾上朱古力醬
將橙片沾上一些朱
古力醬，放回烘焙
紙，於室溫擺放約
一小時。即可使用。

材料 Ingredients

蛋糕
一份基本朱古力蛋糕麵糊
橙汁(1/2個橙)
15克橙皮乾

香橙牛油糖霜
一份基本牛油糖霜
橙皮(1/2個橙)
1/2茶匙香橙香油
1-2滴橙紅色食用色素

裝飾
開縫法式裱花唧嘴
1個唧袋
香橙果醬
6塊鮮橙片
適量朱古力

For Cupcake
1 portion of Basic Chocolate Cupcake
 Batter
1/2 orange juice
15g dried orange peel

For Orange Buttercream
1 portion of Basic Buttercream
1/2 orange zest
1/2 tsp orange essence
1-2 drops orange food coloring

For the Decoration
French Star Piping Tip
1 piping bag
orange jam (instant sugar free orange
 marmalade)
6 pieces of orange slices
chocolate

製作蛋糕

1 混合橙汁和蛋糕糊
將橙汁倒入預先備好的朱古力蛋糕糊裏。

2 加入橙皮乾
加入橙皮乾，拌勻。

3 入爐烘焙
慢慢倒入蛋糕杯，約2/3滿。用170度烘烤20-30分鐘或至呈金黃色。

4 冷卻
完成後，轉移到散熱架上。等蛋糕完全冷卻後才裝飾。

製作香橙牛油糖霜

混合橙皮和牛油糖霜
將橙皮加入一份預先備好的基本牛油糖霜裏。

加入香橙香油
將香橙香油倒入牛油糖霜裏。

加入橙紅色色素
加入橙紅色食用色素。

打發
打發至鬆軟滑身。

裝飾步驟

填入果醬
在蛋糕中間挖一個小孔，填入香橙果醬。

使用開縫星齒唧嘴
使用開縫法式裱花唧嘴，將香橙牛油糖霜唧上蛋糕面。

放上裝飾品
放一塊鮮橙片，或一塊香橙朱古力片作裝飾。

Cupcake Steps

1 Add 1/2 orange juice onto 1 portion of Basic Chocolate Cupcake Batter.

2 Fold in orange juice.

3 Pour into cupcake cups, 2/3 full. Bake for 20-30 minutes or until golden brown.

4 Transfer to cooling rack. Cool down completely before frosting.

Buttercream Steps

1 Add the orange zest into 1 portion of Basic Buttercream.

2 Add in the orange essence.

3 Add in the orange food coloring.

4 Beat until smooth and creamy.

Decoration Steps

1 Cut a hole in each cupcake. Fill it with orange jam.

2 Place the French Star Piping Tip into a piping bag. Place the Orange Buttercream into the piping bag and pipe onto each cupcake.

3 Decorate with a piece of fresh orange slice or chocolate-coat dried orange slices.

TIPS: How to make chocolate-coat dried orange slices?

Steps: 1 Cut orange into 3mm thin slices. Put orange slices on baking tray lined with parchment paper.

2 Bake orange slices in oven at low temp 80°C for 1.5 to 2 hours until water dried out.

3 Melt some dark chocolate in a bowl.

4 Dipped the dried orange slices into melt chocolate and put on parchment paper, let set in room temperature for one hour.

法國

FRANCE

法國是浪漫之都。除了人浪漫,就連食物的文化都一樣浪漫。因為浪漫是需要要時間和心思來營造的。吃一頓法國餐要幾個小時,品嚐紅酒白酒要時間,做法國甜品(比如馬卡龍也需要很長時間)⋯⋯可以說是慢工出細活!法國人很有心思,吃東西除了食物要好吃,環境、情調一樣很重要。烹調食物的時候,也會用很多不同的程序來達到這個目的,所以法式甜品特別精巧繁複。

好友
陳茵媺
的美食印象

自己對法國這浪漫之國一直有份特別情懷,從人到建築設計到時尚品味,它都擁有一種獨樹一幟的傲慢。法國人那份澎湃熱情的愛,彷彿彌漫在整個城市的空氣中似的;這種愛,有時能透過當地特色的甜品感受到:高貴大方的陳列、細緻的味道設計,精美絕倫的手工⋯⋯都在傳遞這份法式的浪漫!每每品嚐法式甜品,我都不自禁地想把時間暫時停止,好去細味這浪漫之都的點滴!

陳茵媺

FRANCE
法國

薰衣草蛋糕

Lavender Cupcakes

我最喜歡的顏色是紫色，而薰衣草就是這種顏色。
薰衣草顏色美，香味也很獨特。很多人會用薰衣草做香薰油，
因為它有紓緩的功效。我收過最特別的薰衣草禮物是一位嫁
給了法國人的好朋友送的薰衣草香包。
她跟我說這些香包是她老公家鄉的特產，
只要將香包放在衣櫃裏，
整個空間就會瀰漫着沁人心田的香氣。這款薰衣草 cupcake，
選用了薰衣草乾和薰衣草食用香油做成。
希望大家在吃完之後，會有一種放鬆心情的感覺。

蛋糕
一份基本雲呢拿牛油蛋糕麵糊
10毫升薰衣草水（先用熱水泡焗薰衣草）
2 茶匙薰衣草乾
1/2 茶匙薰衣草香油（可選）
1-2滴紫色食用色素（可選）

薰衣草牛油糖霜
一份基本牛油糖霜
1/2 茶匙薰衣草香油
1 滴紫色食用色素

燒飾
圓口裱花唧嘴
1個唧袋
迷你馬卡龍
薰衣草乾葉碎
巴黎鐵塔紙牌裝飾

For Cupcake
1 portion of Basic Vanilla Cupcake Batter
10ml lavender water (dried lavender dissolved in hot water)
2 teaspoon dried lavender
l teaspoon lavender essence (optional)
1-2 drops purple food coloring (optional)

For Lavender Buttercream
1 portion of Basic Buttercream
1/2 teaspoon lavender essence
1 drop of purple food coloring

For Decoration
Round Piping Tip
1 piping bag
Mini Macaroon
dried lavender
Eiffel Towel paper topper

Cupcake Steps

1 Add the lavender water little by little into 1 portion of Vanilla Cupcakes.

2 Stir in dried lavender. Add in the lavender essence (optional). Add in a drop of purple food coloring (optional).

3 Pour into cupcake cups, 2/3 full. Bake for 20-30 minutes or until golden brown.

4 Transfer to cooling rack. Cool down completely before frosting.

Lavender Buttercream Steps

1 Add lavender essence to 1 portion of Basic Buttercream.

2 Add in the purple food coloring.

3 Mix until well blended.

Decoration Steps

1 Place the Round Piping Tip into a piping bag. Place the Lavender Buttercream into the piping bag and pipe onto each cupcake. Smooth out with hand spatula.

2 Sprinkle dried lavender on top of each cupcake.

3 Topped with Macaroon.

4 Place an Effiel Tower paper topper as decoration.

製作蛋糕

1 混合薰衣草水和蛋糕糊
將薰衣草水放入一份預先備好的基本雲呢拿牛油蛋糕麵糊裏。

2 加入其他材料
加入薰衣草乾葉碎、薰衣草香油和紫色食用色素。

3 入爐烘焙
慢慢倒入蛋糕杯，約2/3滿。用170度烘烤20-30分鐘或至金黃。

4 冷卻
完成後，轉移到散熱架上。等蛋糕完全冷卻後才裝飾。

製作薰衣草牛油糖霜

加入薰衣草香油
將薰衣草香油加入預先備好的牛油糖霜裏。

加入色素
加入紫紅色食用顏色，攪拌至牛油糖霜呈淡紫色。

打發
將糖霜打發至鬆軟滑身。

裝飾步驟

使用圓口裱花唧嘴
拿1個唧袋，放入圓口裱花唧嘴。將薰衣草牛油糖霜加入，往下推實，紮緊袋口，在蛋糕表面唧上一層糖霜，再用抹刀抹平。

灑上薰衣草乾葉碎
在糖霜上灑上薰衣草乾葉碎。

放上馬卡龍
在每個蛋糕上面放一個迷你馬卡龍甜點作裝飾。

插上裝飾牌
最後，在蛋糕上插一個巴黎鐵塔紙牌作裝飾。

洋梨夏洛蒂蛋糕

Pear Charlotte Cupcakes

Charlotte是法國一種甜品的名稱。通常都是用一般牛油蛋糕，
加入吉士或果醬，然後放上新鮮的生果。
我喜歡啤梨charlotte，因為啤梨的質感好滑身，與吉士是最佳配搭。
這個甜品另一個特別之處就是它的蛋糕由手指餅圍繞着。

浪漫
法式風情~

材料 Ingredients

蛋糕
一份基本雲呢拿牛油麵糊

裝飾
100毫升淡忌廉
適量即食吉士奶黃醬
1包手指餅乾
1個啤梨（切成片）
半吋闊絲帶

For Cupcake
1 portion of Basic Vanilla Cupcake Batter

For Decoration
100ml whipped cream
devon custard
1 packet of lady finger biscuits
1 pear (cut into slices)
1/2 inches width ribbon tape

製作步驟

1 入爐烘焙
在馬芬杯裏放紙模，倒入麵糊，約2/3滿。用170度烘烤20~30分鐘或至呈金黃色。

2 填入吉士奶黃醬
在每一個蛋糕中間挖一個小孔，填入吉士奶黃醬。

3 打發淡忌廉
將淡忌廉打發企身。

4 塗抹忌廉
將忌廉抹上蛋糕面及側面。

5 用手指餅乾圍住蛋糕
用手指餅乾圍住整個蛋糕。用一條絲帶蝴蝶結圍繞手指餅作裝飾。

6 放上啤梨片
在中間放一些啤梨片。

7 製造鏡面效果（可選）
最後，在蛋糕面上掃上果膠即成。

Steps

1. Pour Basic Vanilla Cupcake Batter into cupcake cups, 2/3 full. Bake for 20-30 minutes or until golden brown.
2. Cut a hole in each cupcake. Fill each with devon custard.
3. Whip the cream until peak.
4. Spread the whipped cream on the sides of each cupcake.
5. Cut the lady fingers in half and place around each cupcake. Wrap a ribbon around the lady fingers.
6. Place the pear slices in the middle.
7. Topped with a sugar glaze on top of the pears (optional).

小技巧 Tips

如何自製果膠？

1 糖和1/2杯果汁放入小鍋中，以中火煮熱。

2 將玉米粉倒入剩下的1/2杯果汁中，加入小鍋中一起煮。煮至果膠變濃。

材料 Ingredients

1/2 杯糖
1杯果汁
2湯匙玉米粉

1/2 cup sugar
1 cup fruit juice
2 tablespoons cornstarch

*注意：你所選的果汁會影響到果膠的顏色和味道。

Tips: How to make Sugar Fruit Glaze?

Steps: 1 Put the sugar and 1/2 cup of fruit juice in a saucepan and bring to boil.

2 Dissolve the cornstarch into the remaining 1/2 cup of fruit juice and pour it into the saucepan mixture. Cook until solution thickens.

* The type of fruit juice you choose will affect the flavor and the color of the glaze.

焦糖布甸蛋糕

Crème Brulee Cupcakes

傳統的焦糖布甸比較 creamy，很多人會覺得太甜，
而且有些人喜歡蛋味較濃的味道，
所以不少布甸慢慢演變得像燉蛋。
焦糖布甸的重點當然就在布甸的表面，
燒出來的焦糖，不僅外表美觀，口感也很特別。
我做的這個焦糖布甸 cupcake 保留了脆口的焦糖 topping，
同時蛋糕的質感與傳統布甸又有很大不同，
相信一定能給你們帶來全新的體驗。

材料 Ingredients

蛋糕
一份基本雲呢拿牛油蛋糕麵糊

裝飾
吉士布甸（可自製或在超市買即
　用布甸醬）
白砂糖
鮮草莓
烘焙用手燒槍

For Cupcake
1 portion of Basic Vanilla Cupcake
　Batter

For Decoration
custard pudding (can buy instant
　custard pudding from supermarket)
sugar
fresh strawberries
hand-gas-torch

1 填入布甸
在每一個蛋糕中間挖一個
大孔，填入吉士布甸。

2 撒上白砂糖
撒一些白砂糖在布甸上面。

3 使用火槍
用火槍燒砂糖表面，至呈現
少許燒焦效果。

4 放上鮮草莓
放上鮮草莓作裝飾。

Steps

1　Cut a big hole in the surface of the cupcake. Frost the cake top with a flat layer of custard pudding.

2　Sprinkle some sugar on top.

3　Use hand gas-torch to caramelized the sugar.

4　Decorate with a fresh strawberry.

玫瑰蛋糕
Rose Cupcakes

「玫瑰人生」(La vie en rose) 是法國最著名的香頌，
玫瑰也是法國浪漫精神的象征。玫瑰花可以用作裝飾，
送給愛人表達情意，也可以食用。
在法國到處都可以買到的玫瑰花瓣醬 confit，
是將乾的玫瑰花瓣加入糖做成的。
這種花醬在香港較難買到，不過我用玫瑰茶粉替代，
再用玫瑰牛油糖霜唧出花瓣作裝飾，
做出這個玫瑰 cupcake。
相信會是一份很特別的情人節禮物！

材料 Ingredients

蛋糕

一份基本雲呢拿牛油蛋糕麵糊
10毫升玫瑰茶（預先泡焗）
1茶匙玫瑰茶香粉
1-2滴粉紅色食用
色素（可選）

玫瑰牛油糖霜

一份基本牛油糖霜
2茶匙玫瑰茶香粉
1-2滴粉紅色食用顏色

裝飾

開縫星齒唧嘴
1個唧袋

For Cupcakes

1 portion of Basic Vanilla Cupcake Batter
10ml rose tea (dissolved in hot water)
1 teaspoon rose tea powder
1-2 drops of pink food coloring (optional)

For Rose Buttercream

1 portion of Basic Buttercream
2 teaspoon rose-tea powder
1-2 drops of pink food coloring

For Decoration

Open Star Piping Tip
1 piping bag

蛋糕步驟

1 加入玫瑰茶
將玫瑰茶加入一份預先備好的基本雲呢拿牛油蛋糕麵糊裏。

2 加入玫瑰香粉
加入玫瑰茶香粉。

3 入爐烘焙
慢慢倒入蛋糕杯，約 2/3 滿。用170度烘烤20-30分鐘或至蛋糕表面呈金黃色。

4 冷卻
完成後，轉移到散熱架上。等蛋糕完全冷卻後才裝飾。

玫瑰牛油糖霜步驟

加入玫瑰茶香粉
將玫瑰茶香粉加入預先備好的牛油糖霜裏面。 ➡

加入粉紅色色素
加入粉紅色食用色素。 ➡

打發
打發至鬆軟滑身。

裝飾步驟

使用開縫星齒唧嘴
拿1個唧袋，放入開縫星齒唧嘴。將玫瑰牛油糖霜加入唧袋，往下推實，紮緊袋口，在蛋糕表面唧上花紋。

唧出玫瑰花紋
由中心開始向外繞圈，呈玫瑰花紋狀，完成。

Cupcake Steps

1 Add the rose tea into 1 portion of Basic Vanilla Cupcake Batter.
2 Stir in rose tea powder.
3 Pour into cupcake cups, 2/3 full. Bake for 20-30 minutes or until golden brown.
4 Transfer to cooling rack. Cool down completely before frosting.

Rose Buttercream Steps

1 Add the rose tea powder to 1 portion of Basic Buttercream.
2 Add in the pink food coloring.
3 Mix until well blended.

Decoration steps

1 Place the Open Star Piping Tip into a piping bag. Place the Rose Buttercream into the piping bag and pipe onto each cupcake.
2 Pipe from centre out and around to get a simple swirl rose effect.

意大利
ITALY

意大利是一個文明古國，歷史可追溯至大約 200,000 年前的舊石器時代。許多意大利地區都有重要的考古遺址，有許多值得觀賞的建築物、藝術品。意大利美食種類很多，有披薩、意粉、香腸、海鮮、芝士等等，光是意粉和芝士已有超過 300 種，而且不同城市所做的意菜口味都不一樣！不過令我印象最深的還是意大利的人情味。很多去過意大利的朋友都不懂意大利文，但總會被意大利人的熱情所打動。而如果你問一個意大利人哪裏的意大利菜最好吃？他一定會說是媽媽在家煮的菜。

好友
曹敏莉
的美食印象

我很喜歡意大利這個國家！這裏有我欣賞的建築物，熱情的意大利人及我很喜歡吃的甜品。Tiramisu 是六十年代開始流傳於意大利的蛋糕，自始聞名天下。當吃完重量級的 pizza 及意大利粉，再來個帶濃郁咖啡味道的 Tiramisu 就最適合不過。而我會趁機多飲一杯意大利咖啡，感覺相當幸福！

曹敏莉

ITALY
意大利

提拉米蘇蛋糕

Tiramisu Cupcakes

每一間餐廳做的Tiramisu味道雖然差不多，
但層次可能有些許不同。因為有些喜歡加入很多忌廉，
有些會有蛋糕部分，有些會有芝士部分。
不論層次多與少，濃郁的咖啡味道絕對少不了。
我做的這款Tiramisu味道的cupcake，保留了
傳統做法中的咖啡味和手指餅，裝飾上則以簡約為主。
Tiramisu在意大利語中的意思是「把我拿起」，
希望大家做完這個tiramisu cupcake，
都會很快拿起吃吧！

材料 Ingredients

蛋糕

45克無鹽牛油(室溫)
70克白砂糖
1隻雞蛋
90克自發粉
30毫升脫脂牛奶
20毫升咖啡(先溶於20毫升熱水)
20克植物油
1/2茶匙咖啡香油(可選)

裝飾

圓口裱花唧嘴
1個唧袋
250毫升淡忌廉(或可用即食義大利乳酪芝士)
無糖可可粉(撒面用)
手指餅乾(可選)

For Cupcake

45g unsalted butter (room temperature)
70g caster sugar
1 large egg
90g self-raising flour
30ml skimmed milk
20ml coffee (dissolved in 20ml hot water)
20g vegetable oil
1/2 teaspoon coffee essence (optional)

For Decoration

Round Piping Tip
1 piping bag
250 ml whipped cream (or use instant mascorpone cheese)
cocoa powder (sprinkle on top)
lady fingers (optional)

Cupcake Steps

1 Preheat the oven to 170 degrees Celsius.
2 Beat the butter and sugar until smooth.
3 Add in the egg and mix well.
4 Sift the self-raising flour. Add in the flour little by little to the batter.
5 Mix the milk and coffee and add in the batter little by little. Add in the vegetable oil and coffee essence.
6 Pour into cupcake cups, 2/3 full. Bake for 20-30 minutes or until golden brown.
7 Transfer to cooling rack, and cool down completely before frosting.

Decoration Steps

1 Whip the cream (or Mascorpone cheese - can use immediately and pipe on cakes. Do not need to whip or beat to form cream).
2 Place the Round Piping Tip into a piping bag. Place the Whipped Cream or Mascorpone Cheese into the piping bag and pipe onto each cupcake.
3 Sprinkle cocoa powder on top.
4 Topped with lady fingers (optional).

製作蛋糕

1 準備
將焗爐預熱至攝氏 170 度。牛油置於室溫中回軟。

2 打發牛油
加入白砂糖，和牛油一起打發，至顏色變淺，呈軟滑狀。

3 加入雞蛋
加入打散的蛋液，混合攪拌，直至和牛油融合。

4 篩入粉類
將自發麵粉過篩，然後分3次加入。用切拌法快速拌勻。

5 加入其他材料
將脫脂牛奶跟咖啡液拌勻，逐少加入麵糊中。再加入植物油和咖啡香油。

6 入爐烘焙
慢慢倒入蛋糕杯，約 2/3 滿。用170度烘烤20-30分鐘或至蛋糕表面呈金黃色。

7 冷卻
完成後，轉移到散熱架上。等蛋糕完全冷卻後才裝飾。

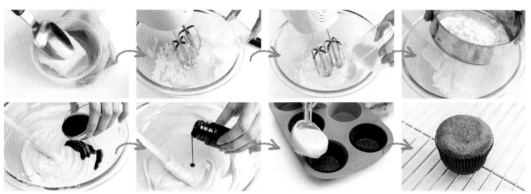

裝飾步驟

打發淡忌廉
將淡忌廉打至企身。

使用圓口裱花唧嘴
拿1個唧袋，放入圓口裱花唧嘴。將淡忌廉加入唧袋，往下推實，紮緊袋口，在蛋糕表面唧上花紋。

撒可可粉
在蛋糕面上撒上無糖可可粉。

放上手指餅
最後，可放上一塊手指餅，再撒上可可粉和食用金箔作裝飾。

成功！

聖誕蛋糕
Panettone Cupcakes

意大利的聖誕蛋糕源自米蘭。意大利人會在聖誕節或者新年期間
吃這種蛋糕。不過近年全世界，甚至連香港，都很容易在超市
買到這種聖誕蛋糕。這種蛋糕的做法很簡單，
就是在麵包中加入乾果。不過通常麵包比較乾身，
所以我做了一個蛋糕版本。
大家在聖誕的時候可以考慮做這款cupcake送給朋友，
他們一定會很開心！

材料 Ingredients

蛋糕
1/2杯乾果
1份基本雲呢拿牛油蛋糕麵糊

For Cupcake
1 portion of Basic Vanilla Cupcake
  Batter
1/2 cup of dried fruits

製作步驟

1 放入乾果
將乾果放入一份預先備好
的基本雲呢拿牛油蛋糕麵
糊裏。

2 入爐烘焙
慢慢倒入蛋糕杯，約2/3
滿。用170度烘烤20-30分
鐘或至蛋糕表面呈金黃色。

3 冷卻
完成後，轉移到散熱架上。

放入乾果。

Steps

1 Add the dried fruits into 1 portion of Basic Vanilla Cupcake Batter.

2 Pour into cupcake cups, 2/3 full. Bake for 20-30 minutes or until golden brown.

3 Transfer to cooling rack.

栗子蛋糕
Mont Blanc Cupcakes

Mont Blanc 是一款經典的意式甜品，它的名字
來自一座叫 Mont Blanc 的雪山，因此這個蛋糕的外形
也很像一座山。我把 Mont Blanc 變成 cupcake，
重點是要用栗子茸唧出一座小山，
然後再灑上糖粉，就像頂着白雪的山頂。
有一個小秘訣就是要做成這個像山的
栗子蓉，一定要用多孔小圓口唧咀！
如果想錦上添花的話，
就可以加一些食用金箔在上面。

材料 Ingredients

蛋糕

1份基本雲呢拿牛油蛋糕麵糊
100毫升淡忌廉

裝飾

圓口裱花唧嘴
多孔小圓口裱花
　唧嘴
1個唧袋
1/2罐即食栗子泥
適量糖霜粉（撒面裝飾用）

For Cupcake

1 portion of Basic Vanilla Cupcake
　Batter
100ml whipped cream

For Decoration

Round Piping Tip
Multi Small Round
　Dots Piping Tip
2 piping bags
1/2 can instant
　Chestnut Purree
icing sugar (for dusting)

製作蛋糕

1 烘焙

準備一份基本雲呢拿牛油
蛋糕麵糊，將其倒入蛋糕
杯，約2/3滿。用170度烘
烤20-30分鐘或至蛋糕表面
呈金黃色。

2 打發淡忌廉

將淡忌廉打企身。

3 填入忌廉

在蛋糕中間挖一個小孔，用
忌廉填滿小孔。

裝飾步驟

使用圓口裱花唧嘴

拿1個嘅袋，放入
圓口裱花唧嘴。在
蛋糕中央唧一個圓
錐體。

使用多孔小圓口裱花唧嘴

再拿1個唧袋，放入多孔小圓口裱花唧嘴。將栗子泥加入唧袋，
往下推實，紮緊袋口，先在蛋糕中央唧一個椎體，再繞圈式地唧
上花紋。

撒上糖霜粉

最後，在蛋糕面上撒糖霜粉作裝飾。

Cupcake Steps

1　Prepare 1 portion of Basic Vanilla Cupcake Batter.
　Pour into cupcake cups, 2/3 full. Bake for 20-30
　minutes or until golden brown.

2　Whip the cream until peak.

3　Cut a hole in each cupcake. Fill center with whipped
　cream.

Decoration Steps

1　Place the Round Piping Tip into a piping bag and start
　to pipe out a small cone in the centre of cupcake.

2　Place another Multi Small Round Dots Piping Tip into a
　piping bag. Place the Chestnut Purree into the piping
　bag and pipe around the centre cone to form a coil.

3　Dusted with icing sugar.

金莎朱古力蛋糕

Ferrero Rocher Cupcakes

在1982年，意大利糖果廠商費列羅推出金莎朱古力，
立即大受歡迎，成為逢年過節送禮的首選！如果大家想特別一點，
不妨試試做一個金莎朱古力cupcake送給別人。
這cupcake裏面有一粒金莎朱古力，頂部也有一粒，
這樣大家每一咬都可以食到金莎啦！喜歡的話
還可以再加一些果仁碎！

雙重
享受！

放入金莎朱古力作餡。

材料 Ingredients

蛋糕

一份基本朱古力蛋糕麵糊
6個金莎朱古力

裝飾

開縫星齒唧嘴
1個唧袋
一份朱古力醬
6個金莎朱古力
果仁（可選）

For the cupcake

1 portion of Basic Chocolate Cupcake
  Batter
6 Ferrero Rocher Chocolate

For the Decoration

Open Star Piping Tip
1 piping bag
1 portion of Chocolate Ganache
6 Ferrero Rocher chocolate as
  decoration
nuts (optional)

製作蛋糕

1 倒入麵糊

將少許基本朱古力蛋糕麵糊倒入蛋糕杯。

2 放入金莎朱古力作餡

將金莎朱古力放在蛋糕杯的中間。

3 再倒入麵糊

將剩餘朱古力蛋糕麵糊倒入蛋糕杯，2/3滿。

4 烘烤並冷卻

用170度烘烤20-30分鐘或至蛋糕表面呈金黃。完成後，轉移到散熱架上。等蛋糕完全冷卻後才裝飾。

裝飾製作

使用開縫星齒唧嘴

拿1個唧袋，放入開縫星齒唧嘴。將朱古力醬加入唧袋，往下推實，紮緊袋口，在蛋糕表面唧上花紋。

放上金莎朱古力作裝飾

在蛋糕面上放一粒金莎朱古力作裝飾。

撒上果仁

最後，可選擇喜歡的果仁撒在蛋糕面上。

Cupcake Steps

1 Pour a little of the Basic Chocolate Cupcake Batter into cupcake cups.

2 Place a Ferrero Rocher chocolate in the middle of each cupcake cups.

3 Continue to pour the remaining cupcake batter into cupcake cups, 2/3 full, until cover up the Ferrero Rocher chocolate.

4 Bake for 20-30 minutes or until golden brown. Transfer to cooling rack. Cool down completely before frosting.

Decoration Steps

1 Place the Open Star Piping Tip into a piping bag. Place the Chocolate Ganache into the piping bag and pipe onto each cupcake.

2 Decorate with Ferrero Rocher chocolate on top.

3 Topped with chopped nuts (optional).

比利時
BELGIUM

比利時是世界公認的朱古力之都，最有名的朱古力生產商大部分來自比利時。在 19 世紀初，比利時朱古力主要供給上流社會，而現時比利時已有超過 2000 間朱古力店，朱古力變成普羅大眾都可以隨時買到的甜食。不過，因為朱古力越來越受歡迎，這幾年可可豆的價格也節節攀升，遲一些不知道朱古力會不會再度成為奢侈的食品呢？除了朱古力外，比利時的窩夫、青口、薯條和啤酒都很出名。如果你有機會到比利時，不妨好好品嚐一番！

好友
張嘉兒
的美食印象

我是百分百的 chocoholic，差不多每天都會吃朱古力，不吃好像這天都不完整！雖然一般市面上容易買到的朱古力已經滿足到我的要求，但自認朱古力 maniac，當然好想有一天可以去到全球最有名的朱古力產地試試最頂級的藝術品啦！對，我說的就是比利時！聽說，比利時擁有過百年做朱古力的歷史，直到現在他們的製作過程還保留了傳統工藝！更厲害的是，他們選用的是當地的新鮮食材，可以直接將可可從可可豆提煉出來立即做成朱古力，減少化學添加成分！正所謂新鮮的材料是最無價的，加上比利時師傅的經驗，一定是頂級美食！哎！好想現在就飛去比利時試試又滑又香又甜的比利時朱古力啊！

比利時窩夫蛋糕

Belgium Waffle Cupcakes

在比利時的大街小巷都可以買到窩夫，甚至有流動窩夫車，
可以說是當地的國民美食！為了製作有窩夫特色的蛋糕，
我選擇了用雪糕窩夫甜筒來做材料。窩夫甜筒的好處，
就是可以跟蛋糕一起放進焗爐焗烤，
做出來的效果也很可愛，
就好像一個真正的雪糕一樣。

材料 Ingredients

蛋糕

90克無鹽牛油（室溫）
140克 白砂糖
2隻雞蛋
180克自發麵粉
80毫升脫脂牛奶（室溫）
40克植物油
1茶匙香草精油
6個窩夫雪糕筒（超市可買到）

裝飾

開縫星齒唧嘴
1個唧袋
一份蛋白霜
彩色朱古力糖珠
朱古力淋面醬
窩夫（蛋糕面裝飾用）

For Cupcake

90g unsalted butter, at room temperature
140g caster sugar
2 large eggs
180g self-raising flour
80ml skimmed milk, at room temperature
40g vegetable oil
1 teaspoon vanilla extract
6 waffle ice-cream cones (can buy in
　supermarket)

For the Decoration

Open Star Piping Tip
1 piping bag
1 portion of Meringue Frosting
rainbow chocolate sprinkles
chocolate sauce
waffle (put on the side for decoration)

雪糕筒
在超市就
買得到！

Cupcake Steps

1 Preheat the oven to 170 degrees Celsius.

2 Beat the butter and sugar until smooth.

3 Add in the egg and mix well.

4 Sift the self-raising flour. Add in the flour little by little to
　the batter.

5 Add in the milk little by little. Add in the vegetable oil.
　Add in the vanilla essence.

6 Pour into waffle cones, 2/3 full. Bake for 20-30
　minutes or until golden brown.

7 Transfer to cooling rack. Cool down completely before
　frosting

Decoration Steps

1 Place the Open Star Piping Tip into a piping bag.
　Place the Meringue Frosting into the piping bag and
　pipe onto each cupcake.

2 Drizzle some chocolate sauce on top of meringue
　cream. Sprinkle rainbow chocolate sprinkles on top of
　cupcakes.

3 Place a small piece of waffle or cookie on the side of
　cupcake.

製作蛋糕

1 準備
將焗爐預熱至攝氏170度。牛油在室溫中回軟。

2 打發牛油
加入白砂糖,和牛油一起打發,直到顏色變淺,呈軟滑狀。

3 加入雞蛋
加入打散的蛋液,混合攪拌,直至和牛油融合。

4 篩入粉類
將自發麵粉過篩,然後分3次加入。用切拌法快速拌勻。

5 加入其他材料
加入脫脂牛奶、植物油、香草精油,攪拌均勻。

6 入爐烘焙
倒入雪糕筒裏面,呈2/3滿。用170度烘烤20-30分鐘或至蛋糕表面呈金黃色。

7 冷卻
完成後,轉移到散熱架上。等蛋糕完全冷卻後才裝飾。

裝飾步驟

使用開縫星齒唧嘴
拿1個唧袋,放入開縫星齒唧嘴。將蛋白糖霜加入唧袋,往下推實,紮緊袋口,在蛋糕表面唧上花紋。

撒上糖珠
淋上朱古力醬,再撒上彩色朱古力糖珠。

插一塊窩夫作裝飾
插一塊窩夫在蛋糕上作裝飾。

完成!

比利時曲奇蛋糕

Speculoos Cookies Cupcakes

Speculoos是比利時傳統的曲奇餅。餅乾裏面有杏仁粉，也有很多不同的香料。而我第一次吃這個餅乾是在比利時的朱古力博物館。店員給我們一塊沾了朱古力的Speculoos！味道好香，好特別！我想把這味道保留下來，就做了這個cupcake。如果大家有機會去比利時的話，除了買朱古力之外，可以買一罐 Speculoos麵包醬，包你一試難忘！

材料 Ingredients

曲奇餅

250克普通麵粉
185克紅糖
125克無鹽牛油
1/4湯匙肉桂粉
1/2湯匙混合香草粉
1隻雞蛋
50克杏仁粉(幼磨)

脆餅底

40克無鹽牛油(溶化)
80克比利時曲奇餅(壓碎)

蛋糕

45克無鹽牛油
70克紅糖
1隻雞蛋
75克自發麵粉
15克杏仁粉
1/2茶匙肉桂粉
1茶匙混合香草粉
40毫升脫脂牛奶
40克植物油

裝飾

開縫星齒唧嘴
1個唧袋
一份朱古力醬
6個比利時曲奇餅
杏仁碎

For Speculoos Cookie

250g all purpose flour
185g brown sugar
125g unsalted butter
1/4 tablespoon grounded cinnamon
1/2 tablespoon all spices (powder of mixed ginger, nutmeg, etc)
1 large egg
50g almond powder

Bottom layer

40g unsalted butter (melted)
80g speculoos cookies (crumbs)

Cupcake Layer

45g unsalted butter
70g brown sugar
1 large egg
75g self-raising flour
15g almond powder
1/2 teaspoon grounded cinnamon
1 teaspoon of all spices powder
40ml skimmed milk
20g vegetable oil

Decoration

Open Star Piping Tip
1 piping bag
1 portion of Chocolate ganache
6 speculoos cookies
chopped almond

製作曲奇餅

1 準備
將焗爐預熱至攝氏160度。
牛油置於室溫中回軟。

2 製作麵糰
混合所有材料，直到拌勻成
麵糰。

3 冷藏
用保鮮紙包好麵糰放入雪
櫃大概35分鐘。

4 壓平
將麵糰用擀麵棍壓平，推薄
至約4mm-5mm厚。

5 切模
用曲奇切模，切出不同所需
形狀。

6 入爐烘焙
放入焗爐烘烤15分鐘或至
金黃。

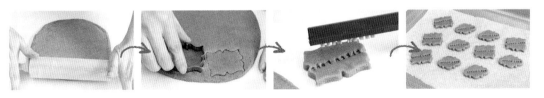

製作蛋糕

1 製作蛋糕脆餅底層
將牛油溶液加入餅乾碎，混
勻。將餅乾碎放入蛋糕杯
底，壓平至4毫米厚。

2 預熱
將焗爐預熱至攝氏170度。

3 打發牛油
分3次加入砂糖，將牛油和
糖混合打發，直到顏色變
淺、呈軟滑狀。

4 加入雞蛋
加入雞蛋，攪拌均勻。

5 篩入粉類
將自發麵粉、杏仁粉、肉桂
粉和香草粉混合過篩，然後
逐少加入雞蛋牛油溶液中。

6 加入其他材料
逐少加入牛奶，再加入植物
油。

7 入爐烘焙
倒入蛋糕紙杯中，呈2/3
滿。用170度烘烤20-30分
鐘或至蛋糕表面呈金黃色。

8 冷卻
完成後，轉移到散熱架上。
等蛋糕完全冷卻後才裝飾。

裝飾步驟

使用開縫星齒唧嘴
拿1個唧袋，放入開縫星齒唧嘴。將朱古力醬放入唧
袋，往下推實，紮緊袋口，在蛋糕表面唧上花紋。 ➡

放上比利時曲奇餅
放一塊比利時曲奇餅在蛋糕面作裝
飾。再撒上杏仁碎。

Cookies Steps

1 Preheat the oven to 160 degrees celcius.

2 Mix all the ingredients together to form dough.

3 Put dough in the fridge for 35 minutes.

4 Roll out the dough 4-5mm thick.

5 Cut out shapes with a cookie cutter.

6 Bake for 15 minutes or until golden brown.

Cupcake Steps

1 For bottom crust: Melt unsalted butter, mix in speculoos cookies into crumbs. Press firmly on bottom of each cupcake liners about 4mm thick.

2 For cupcake: Preheat the oven to 170 degrees Celsius.

3 Beat the butter and sugar until smooth.

4 Add in the egg and mix well.

5 Add the self-raising flour, cinnamon, all spice and almond powder mixture into batter.

6 Add in the milk little by little. And add in the vegetable oil.

7 Pour batter into cupcake cups on top of cookie crust, 2/3 full. Bake for 20-30 minutes or until golden brown.

8 Transfer to cooling rack. Cool down completely before frosting.

Decoration Steps

1 Place the Open Star Piping Tip into a piping bag. Place the Chocolate Ganache into the piping bag and pipe onto each cupcake.

2 Decorate with a speculoos cookie on top of each cupcake. Sprinkle with chopped almond.

黑朱古力蛋糕

Dark Chocolate Cupcakes

代表比利時的味道怎麼能少了傳統的朱古力蛋糕！為了令蛋糕有更加濃烈的朱古力味，
我特別加入朱古力粒，而且蛋糕的頂部是選用朱古力和淡忌廉製成的朱古力醬，
最後用朱古力塊裝飾。這款cupcake絕對是朱古力迷的摯愛！

材料 Ingredients

蛋糕
1份基本朱古力蛋糕麵糊
15克黑朱古力粒

裝飾
多瓣花齒唧嘴
1個唧袋
一份朱古力醬
黑朱古力碎
無糖可可粉
食用銀色糖珠

For Cupcake
1 portion of Basic Chocolate Cupcake
    Batter
15g dark chocolate chips

For Decoration
Closed Star Piping Tip
1 piping bag
1 portion of Chocolate Ganache
dark chocolate flakes
cocoa powder
edible silver sugar beads

製作蛋糕

1 放入黑朱古力
將黑朱古力粒放入一份預先備好的基本朱古力蛋糕糊裏面。

2 入爐烘焙
慢慢倒入蛋糕杯，約2/3滿。用170度烘烤20-30分鐘或至金黃。

3 冷卻
完成後，轉移到散熱架上。等蛋糕完全冷卻後才裝飾。

裝飾步驟

使用多瓣花齒唧嘴
拿1個唧袋，放入多瓣花齒唧嘴。將朱古力醬放入唧袋，往下推實，紮緊袋口，在蛋糕表面唧上花紋。

放上黑朱古力碎
在朱古力醬上放上一些黑朱古力碎作裝飾。

撒上無糖可可粉和銀色糖珠
最後，撒上無糖可可粉和食用銀色糖珠。

Cookies Steps

1 Add melted dark chocolate chips into 1 portion of Basic Chocolate Cupcake Batter.

2 Pour into cupcake cups, 2/3 full. Bake for 20-30 minutes or until golden brown.

3 Transfer to cooling rack. Cool down completely before frosting.

Cupcake Steps

1 Place the Closed Star Piping Tip into a piping bag. Place the Chocolate Ganache into the piping bag and pipe onto each cupcake.

2 Decorate with chocolate flakes.

3 Sprinkle cocoa powder and edible silver sugar beads on top.

白朱古力蛋糕
White Chocolate Cupcake

有些朋友喜歡吃白朱古力，好像我一樣。我吃了可可粉較多或者濃度比較高的朱古力後，
會不停打噴嚏，所以更喜歡吃白朱古力。在這個cupcake裏，我加了白朱古力粒，
表面唧上白朱古力牛油霜，頂部再放上一塊白朱力片作裝飾，賣相一流。
在3月14的白色情人節，男士們不妨考慮一下做這款cupcake送給你心儀的對象？

材料 Ingredients

蛋糕

一份基本雲呢拿牛油蛋糕麵糊
30克白朱古力(已溶化)
20克白朱古力粒

白朱古力牛油糖霜

100克無鹽牛油
200克糖霜粉
2-3茶匙脫脂牛奶
70克白朱古力(已溶化)
1.5湯匙濃奶油

裝飾

多瓣花齒唧嘴
1個唧袋
白朱古力碎片
6塊白朱力片
食用彩色糖珠

For Cupcake

1 portion of Basic Vanilla Cupcake Batter
30g white chocolate (melted and fold in batter)
20g white chocolate chips

White Chocolate Buttercream

100g unsalted butter
200g icing sugar
2-3 teaspoon of skimmed milk
70g melted white chocolate
1.5 tablespoon double cream

For Decoration

Closed Star Piping Tip
1 piping bag
white chocolate flakes
6 white chocolate
edible sugar beads

製作蛋糕

1 混合白朱古力和蛋糕糊
將白朱古力座熱水溶化後,放入一份預備好的基本雲呢拿牛油蛋糕糊裏面。加入白朱古力粒。

2 入爐烘焙
倒入蛋糕杯,2/3滿。用170度烘烤20-30分鐘或至表面呈金黃。

3 冷卻
完成後,轉移到散熱架上。等蛋糕完全冷卻後才裝飾。

製作白朱古力牛油糖霜

溶化白朱古力
隔熱水座溶白朱古力。

打發牛油
將室溫牛油打發至軟綿。將糖霜粉篩過,然後加入牛油打發至順滑。

加入其他材料
逐少加入脫脂牛奶。加入濃奶油拌勻。最後加入已經溶掉白朱古力漿。

繼續打發
繼續打發至鬆軟滑身。

裝飾步驟

使用多瓣花齒唧嘴
拿1個唧袋,放入多瓣花齒唧嘴。將白朱古力牛油糖霜加入唧袋,往下推實,紮緊袋口,從蛋糕表面唧上花紋。

撒上白朱古力碎片
撒上白朱古力碎片和食用彩色糖珠作裝飾。最後插上一塊白朱力片。

Cupcake Steps

1 Add the melted white chocolate into 1 portion of Basic Vanilla Cupcake Batter. Stir in the white chocolate-chips.

2 Pour into cupcake cups, 2/3 full. Bake for 20-30 minutes or until golden brown.

3 Transfer to cooling rack. Cool down completely before frosting.

Buttercream Steps

1 Melt the chocolate in a small bowl. Beat the softened unsalted butter.

2 Sift the icing sugar in another bowl, and add in the mixture, little by little. Add in the milk, 1 teaspoon at a time.

3 Add in the double cream. Last add in the melted chocolate.

4 Beat until smooth and creamy.

Decoration Steps

1 Pipe the White Chocolate Buttercream on each cupcake using a Closed Star Piping Tip.

2 Sprinkle white chocolate flakes and sugar beads on top. Insert a piece of white chocolate bar on top.

美

UNITED STATES

美國是由 50 個洲和華盛頓哥倫比亞特區所組成，因國土面積大，不同的州區的天氣差異也很大。因為這個原因，食材的選擇也很豐富。很多人形容美國飲食文化是以 comfort food 為主，意思是指某些食品雖然製作簡單但卻能喚起家一般溫馨舒服的感覺。不過不知道是否食物給了舒服感覺的原因，人的胃口也大了，所以美國食物的份量也很大！我特別喜歡美國的小食，比如漢堡、甜甜圈，感覺這裏的食物總是五彩繽紛的，好像當地人一樣樂觀風趣，充滿活力！

好友
曹敏莉
的美食印象

美國的甜品真的比較有份量！

我在美國出生，小時候接觸的蛋糕都特別大 Size，一個人往往很難吃完！

現在的我喜愛的甜品通常香味比較重、甜味比較淡，但唯獨是戒不了從小鐘意的草莓芝士蛋糕：真的好喜歡這種溶在口裏，甜在心裡的感覺！所以每次返回美國都一定會品嚐草莓芝士蛋糕才感到旅程完滿。

曹敏莉

UNITED STATES
美國

紅蘿蔔蛋糕

Carrot Cupcakes

紅蘿蔔蛋糕配上忌廉芝士糖霜被譽為1970年代在美國最受歡迎的五大甜品之一。紅蘿蔔的糖粉很高，所以最初用來代替糖去做蛋糕，因為成本便宜很多。除了便宜外，它還可以令蛋糕更加濕潤、軟熟。再配上酸酸的忌廉芝士糖霜，可以中和蛋糕的甜，味道剛剛好！我喜歡加玉桂粉，味道更香，再加一些核桃，口感更佳。

蛋糕

一份基本雲呢拿牛油蛋糕麵糊
80克生紅蘿蔔(切絲)
20克核桃(可選)
1茶匙肉桂粉

裝飾

圓口裱花唧嘴
1個唧袋
一份忌廉芝士糖霜
一粒大的橙色棉花糖
一粒大的綠色棉花糖
糖做的白兔公仔

For the Cupcake

1 portion of Basic Vanilla Cupcake Batter
80g raw carrots (shredded)
20g chopped walnuts (optional)
1 teaspoon Grounded Cinnamon

For the Decoration

Round Piping Tip
1 piping bag
1 portion of Cream Cheese Frosting
1 large size orange fruity marshmallow
1 large size green fruity marshmallow
1 sugar rabbit (optional)

小貼士
Tips

如棉花糖顏色不夠鮮艷，
可噴上食用色素。

Cupcake Steps

1 Add in grated raw carrots, walnuts (optional) and Grounded Cinnamon into 1 portion of Basic Vanilla Cupcake Batter.

2 Pour into cupcake cups, 2/3 full. Bake for 20-30 minutes or until golden brown

3 Transfer to cooling rack. Cool down completely before frosting.

Decoration Steps

1 Place the Round Star Piping Tip into a piping bag. Place the Cream Cheese Frosting into the piping bag and pipe onto each cupcake.

2 Cut the orange marshmallow into half and place on top of the frosting.

3 Cut the green marshmallow into half and cut some slits to make it like the stem of the carrot.

4 Place the sugar rabbit on the side (optional).

Tips: If the marshmallow's colour is not bright enough, you can also use food colorant.

製作蛋糕

1 混合材料
將生紅蘿蔔、核桃、肉桂粉加入一份預先預備好的基本牛油蛋糕糊裏面。

2 入爐烘焙
慢慢倒入蛋糕杯，約 2/3 滿。用170度烘烤20-30分鐘或至蛋糕表面呈金黃色。

3 冷卻
完成後，轉移到散熱架上。等蛋糕完全冷卻後才裝飾。

裝飾

使用圓口裱花唧嘴
拿1個唧袋，放入圓口裱花唧嘴。將忌廉芝士糖霜加入唧袋，往下推實，紮緊袋口，在蛋糕表面唧上花紋。

放上橙色棉花糖
將橙色棉花糖剪開一半，然後放在忌廉芝士糖霜上面。

放上綠色棉花糖
將綠色棉花糖剪開一半，然後剪成條狀，製成紅蘿蔔的頂部。

放上白兔糖公仔(可選)
最後，放一個糖做的白兔公仔作裝飾。

香蕉蛋糕

Banana Cupcakes

香蕉蛋糕是我的童年回憶！在我小時候，有一個姨姨
常常會焗香蕉蛋糕給我一家吃，她可以說是
我烘焙愛好的啟蒙者！她常常對我們說，
用熟透的香蕉做的香蕉蛋糕是最好吃。而在我家中，一定會
有香蕉出現，因為我爸媽好喜歡吃這種水果。他們說香蕉有
豐富的維他命 B_6、C、B_2，對身體非常好。所以，下次有熟
透的香蕉記得可留作做香蕉蛋糕哦！

香蕉有
豐富維他命！

蛋糕步驟

1 混合香蕉、肉桂和牛油蛋糕糊
將香蕉和肉桂粉加入一份預先備好的基本雲呢拿牛油蛋糕麵糊裏面。

2 入爐烘焙
慢慢倒入蛋糕杯，約2/3滿。用170度烘烤20-30分鐘或至蛋糕表面呈金黃色。

3 冷卻
完成後，轉移到散熱架上。等蛋糕完全冷卻後才裝飾。

裝飾步驟

打發忌廉
將鮮忌廉打企身。

使用開縫星齒唧嘴
拿1個唧袋，放入開縫星齒唧嘴。將鮮忌廉加入唧袋，往下推實，紮緊袋口，在蛋糕表面唧上花紋。

用香蕉片、肉桂粉裝飾
放上香蕉片或香蕉糖、香蕉乾，撒上肉桂粉。

材料 Ingredients

蛋糕

一份基本雲呢拿牛油蛋糕麵糊
50克香蕉(已熟透 / 壓成泥醬)
1茶匙肉桂粉

裝飾

開縫星齒唧嘴
1個唧袋
250毫升淡忌廉
香蕉糖
新鮮香蕉
或香蕉乾
肉桂粉

For Cupcake

1 portion of Basic Vanilla Cupcake Batter
50g bananas (mashed)
1 teaspoon grounded cinnamon

For Decoration

Open Star Piping Tip
1 piping bag
250ml whipped cream
6 pieces of banana candies/fresh banana slices /dried bananas
grounded cinnamon

小貼士 Tips

最好不要在蛋糕中加入太多香蕉，否則會令水分過多，難以發起。

Cupcake Steps

1. Add the bananas and grounded cinnamon into 1 portion of Vanilla Cupcakes Batter.
2. Pour into cupcake cups, 2/3 full. Bake for 20-30 minutes or until golden brown.
3. Transfer to cooling rack. Cool down completely before frosting.

Decoration Steps

1. Whip the cream to peak.
2. Place the Open Star Piping Tip into a piping bag. Place the whipped cream into the piping bag and pipe onto each cupcake.
3. Topped with fresh banana slices, dried bananas or banana candy. Sprinkled with grounded cinnamon on top.

Tips: Do not add too much banana to the batter, or the banana will become very soggy, as there is too much fluid.

草莓芝士蛋糕

Strawberry Cheesecake Cupcakes

美國的芝士蛋糕通常都比較大件，
而芝士蛋糕的底通常用餅乾碎製成。我的這款
草莓芝士cupcake採用牛油蛋糕底，添加果醬和
新鮮草莓，再配上香滑的忌廉芝士糖霜，
口味更加大眾化，
適合抗拒濃味芝士的人食用。

材料 Ingredients

蛋糕

55克無鹽牛油（室溫）
50克白砂糖
1隻雞蛋
100克自發粉
40毫升脫脂牛奶
25毫升植物油
1/2茶匙香草香油
1/2湯匙草莓果醬
5粒鮮草莓（切成小粒果肉）

裝飾

多瓣花齒唧嘴
1個唧袋
一份忌廉芝士糖霜
粉紅色的食用色素
草莓果醬
消化餅乾碎
3粒鮮草莓（切半）

For the cupcake

55g unsalted butter, at room
    temperature
50g caster sugar
1 large egg
100g self-raising flour
40ml skimmed milk
25ml vegetable oil
1/2 teaspoon vanilla essence
1/2 tablespoon strawberry jam
5 pcs strawberries (cut into
    small pcs)

Decoration

Closed Star Piping Tip
1 piping bag
1 portion of Cream Cheese
    Frosting
1-2 drops of pink food coloring
strawberry jam (as filling)
digestive biscuit crumbs (for
    decoration)
3 strawberries (cut into halves)

步驟

1 準備
將焗爐預熱至攝氏170度。
牛油置於室溫中回軟。

2 打發牛油
打發牛油和糖，直到顏色變
淺，呈軟滑狀。

3 加入雞蛋
加入雞蛋，拌勻。

4 篩入粉類
將自發粉篩過，逐少加入。

5 加入其他材料
逐少加入脫脂牛奶、植物
油、香草香油。加入草莓果
醬和草莓粒。

6 入爐烘焙
慢慢倒入蛋糕杯，約2/3
滿。用170度烘烤20-30分
鐘或至蛋糕表面呈金黃色。

7 冷卻
完成後，轉移到散熱架上。
等蛋糕完全冷卻後才裝飾。

加入草莓粒。

裝飾步驟

加入粉紅色素
將粉紅色的食
用色素加入一
份忌廉芝士糖
霜裏。

填入草莓果醬
在蛋糕中間挖
了一小孔，用
草莓果醬填滿
這個孔。

使用多瓣花齒唧嘴
拿1個嘅袋，放入多瓣花齒唧
嘴。將忌廉芝士糖霜加入唧
袋，往下推實，紮緊袋口，在
蛋糕表面唧上花紋。

**用消化餅、鮮草莓
作裝飾**
撒上消化餅乾碎。
最後，放一粒鮮草莓
作裝飾。

Cupcake Steps

1 Preheat the oven to 170 degrees
   Celsius.

2 Beat the butter and sugar until light
   yellow in color and smooth.

3 Add in the egg and mix well.

4 Sift the self-raising flour. Add in the
   flour little by little to the batter.

5 Add in the milk little by little. Add
   in the vegetable oil and the vanilla
   essence. Stir in the Strawberry
   Jam and strawberry bits.

6 Pour into cupcake cups, 2/3 full.
   Bake for 20-30 minutes or until
   golden brown.

7 Transfer to cooling rack. Cool
   down completely before frosting.

Decoration Steps

1 Add 1-2 drops of pink food coloring
   into the Cream Cheese Frosting.

2 Cut a hole in the middle of
   each cupcake. Fill Centre with
   strawberry jam.

3 Place the Closed Star Piping Tip
   into a piping bag. Place the Cream
   Cheese Frosting into the piping
   bag and pipe onto each cupcake.

4 Sprinkle with digestive biscuit
   crumbs. Decorate with half a fresh
   strawberry.

青檸批蛋糕
Key Lime Pie Cupcakes

說到青檸批，一定要說它的起源。有傳說是由一班漁民創造
出來。因為出海期間要帶一些可以儲存比較長時間，
而且含有豐富營養價值的食物，好像罐頭牛奶、煉奶、
青檸、雞蛋等等。而這些食物，就可以做到青檸批。
不過因為船上不會有焗爐，所以傳統的做法是不用焗的。
這次我就想到可以加一些蛋糕的元素
去製成一個軟綿綿的青檸批蛋糕。

軟綿綿的
「青檸批」好特別！

材料 Ingredients

脆餅底
40克無鹽牛油(溶化)
80克消化餅乾碎

蛋糕部分
一份基本雲呢拿牛油蛋糕麵糊
2湯匙青檸汁
1/2個青檸(皮)

青檸奶酪面層
2個蛋黃
200毫升煉奶
1/4杯青檸汁

裝飾
開縫星齒唧嘴
1個唧袋
250毫升淡忌廉
青檸皮碎
青檸片

Bottom Layer
40g unsalted butter, melted
80g digestive biscuits (crumbs)

For Cupcake
1 portion of Basic Vanilla Cupcake Batter
2 tablespoon fresh lime juice
zest of 1/2 lime

Key Lime Layer (top layer)
2 egg yolks
200ml condensed milk
1/4 cup fresh lime juice

For Decoration
Open Star Piping Tip
1 piping bag
250 ml whipping cream
lime zest for decoration
Fresh slices of lime or lime jelly candies

蛋糕步驟

1 製作脆餅底
將牛油加熱溶解,再與餅乾碎混合。將餅乾碎倒入蛋糕杯的底部,壓平至4mm-5mm厚。

2 混合青檸汁、青檸皮和牛油蛋糕麵糊
在預先備好的基本雲呢拿牛油蛋糕麵糊裏加入青檸汁和青檸皮。

3 入爐烘焙
將麵糊慢慢倒入蛋糕杯,約1/2滿。烘烤15-20分鐘,拿出來備用。

4 製作青檸奶酪面層
將蛋黃、煉奶與青檸汁混合,倒在蛋糕面上。再放入焗爐中烘烤10-15分鐘。

5 冷卻
完成後,轉移到散熱架上。等蛋糕完全冷卻後才裝飾。

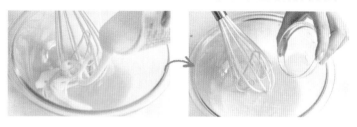

裝飾步驟

打發淡忌廉
將淡忌廉打企身。

使用開縫星齒唧嘴
拿1個唧袋,放入開縫星齒唧嘴。將鮮忌廉加入唧袋,往下推實,紮緊袋口,在蛋糕表面唧上花紋。

用青檸皮、青檸片作裝飾
撒上青檸皮碎。最後,放一塊鮮青檸片作裝飾。

Cupcake Steps

1 For the bottom crust: Melt the unsalted butter, add in digestive cookie crumbs, hand rub until all ingredients are well mixed together. Press cookie crumbs firmly at the bottom of each cupcake liner around 4-5mm thick.

2 For the cupcake: Add in the lime juice and lime zest into 1 portion of Basic Vanilla Cupcake Batter.

3 Pour into cupcake cups, 1/2 full. Bake for 15-20 minutes and take it out set aside.

4 For the Key Lime top Layer: Whisk the egg yolks, condensed milk and lime juice in a bowl until well blended. Pour on top of the cupcakes. Bake for another 10-15 minutes or until the top layer is sturdy.

5 Transfer to cooling rack. Cool down completely before frosting.

Decoration Steps

1 Place the Open Star Piping Tip into a piping bag.

2 Place the Whipped Cream into the piping bag and pipe onto each cupcake.

3 Sprinkle some lime zest on top of each cupcake. Decorate with a fresh slice of lime or lime jelly candies.

加拿大
CANADA

加拿大是我出生的地方，因此我對這裡有很多美好的回憶。加拿大地方大，空氣好，風景美，探訪當地時常常會接觸到大自然，尤其是楓葉。加拿大也是一個多國文化的國家，有來自不同地方的人，所以當地的美食也非常多元化。不過，這個地方的食物通常比較直接，意思是食物的賣相不花巧。烹調時強調食材的新鮮，烹調過程並不複雜。若要選擇代表加拿大的甜品，我會選出那些在我成長中產生影響的，希望透過這些特別的味道，讓大家對我的家鄉有更多了解。

好友
張嘉兒
的美食印象

我在溫哥華長大，所以很容易就可以用到百分百全天然的楓樹糖漿。我非常喜歡這個味道，小時候幾乎每天都用 pancake 來配楓樹糖漿，相信很多小朋友也有一樣感受：楓樹糖漿才是早餐的主角嘛！每一層的 pancake 都被甜甜的楓樹糖漿包裹，再佐以溫哥華特產的新鮮藍莓、黑莓和樹莓，實在是人間美味！

給大家一個健康小 tips：雖然楓樹糖漿比 refined sugar 健康，可以提供一些礦物質，但糖分仍然很高，建議不要學我吃這麼多！反而，pancake 上的雜莓有抗氧化功效，多吃也無妨！

張嘉兒

楓樹糖漿蛋糕

Maple Syrup Cupcakes

在加拿大，楓樹無處不在！加拿大天氣很冷，
楓樹在冬天的時候會製造糖。到了春天，這些糖就可以從楓樹
汁液中提取出來。不過要經過加熱、蒸發水分等
繁複工序後，楓葉糖漿才可食用。幾乎每個家庭都會有一瓶楓
樹糖漿，用來搭配各種美食！我覺得它的
味道很獨特，如果將它加入cupcake一起烘焙，一定會好香。

蛋糕

55克無鹽牛油（室溫）

50克白砂糖

1隻雞蛋

100克自發粉

40毫升脱脂牛奶

25毫升植物油

20毫升楓樹糖漿

楓樹糖漿蛋白奶油糖霜

1份法式蛋白奶油糖霜

2茶匙楓樹糖漿

1-2滴食用色素（可選）

裝飾

圓口裱花唧嘴

1個唧袋

黑朱古力

葉形糖果 / 楓樹糖漿乾碎片（作裝飾用）

For Cupcake

55g unsalted butter (room temperature)

50g caster sugar

1 large egg

100g self-raising flour

40ml skimmed milk

25ml vegetable oil

20ml maple syrup

For Maple Syrup French Meringue Buttercream

1 portion of French Meringue Buttercream

2 teaspoons Maple Syrup

1-2 drops of food coloring (optional)

For Decoration

Round Piping Tip

1 piping bag

dark chocolate (for decoration)

leaf like candies/dried maple syrup bits as decoration

製作蛋糕

1 準備

將焗爐預熱至攝氏170度。牛油置於室溫中回軟。

2 打發牛油

加入白砂糖，和牛油一起打發，至顏色變淺，呈軟滑狀。

3 加入雞蛋

加入打散的蛋液，混合攪拌，直至和牛油融合。

4 篩入粉類

自發麵粉過篩，然後分3次加入。用切拌法快速拌勻。

5 加入其他材料

加入脱脂牛奶、植物油、楓樹糖漿，攪拌均勻。

6 入爐烘焙

慢慢倒入蛋糕杯，約 2/3 滿。用170度烘烤20-30分鐘或至蛋糕表面呈金黃色。

7 冷卻

完成後，轉移到散熱架上。等蛋糕完全冷卻後才裝飾。

製作楓樹糖漿法式蛋白奶油霜

拌入楓樹糖漿
將楓樹糖漿拌勻入預先打發好一份蛋白奶油霜中。

加入食用色素
如需調色，將1-2滴食用色素滴入奶油霜中，再用手抹刀混勻。

裝飾步驟

使用圓口裱花唧嘴
拿1個唧袋，放入圓口裱花唧嘴。放入楓樹糖漿法式蛋白奶油霜，往下推實，紮緊袋口，在蛋糕表面唧上花紋。

用朱古力畫出樹枝
黑朱古力隔熱水座溶，裝入唧袋，在牛油紙上畫出樹枝。可先畫出簡單輪廓，再反復加粗。

撒上糖果
用葉形糖果或楓糖乾碎片作裝飾。

放上奶油霜面
將楓樹置於奶油霜上作裝飾。

Cupcake Steps

1 Preheat the oven to 170 degrees Celsius.

2 Beat the butter and sugar until smooth.

3 Add in the egg and mix well.

4 Sift the self-raising flour. Add in the flour little by little to the batter.

5 Add in the milk little by little. Add in the vegetable oil. Stir in the maple syrup

6 Pour into cupcake cups, 2/3 full. Bake for 20-30 minutes or until golden brown.

7 Transfer to cooling rack. Cool down completely before frosting.

Maple Syrup French Meringue Buttercream Steps

1 Stir in the Maple syrup into 1 portion of French Meringue Buttercream.

2 Add in 1-2 drops of food coloring (optional). Mix until smooth and creamy.

Decoration Steps

1 Place the Round Piping Tip into a piping bag. Place the Maple Syrup French Meringue Buttercream into the piping bag and pipe onto each cupcake.

2 Melt the dark chocolate. Use the melted chocolate and pipe out a tree with only branches onto baking paper. Start with a thin branch, let it dry and add extra layers to thicken it.

3 Decorate the chocolate branches with Leaf like candies/dried maple syrup bits as decoration.

4 Place on top of the buttercream.

加拿大果仁朱古力蛋糕

Nanaimo Bar Cupcakes

Nanaimo bar是加拿大果仁朱古力蛋糕。名字是來自卑詩省一個叫做Nanaimo的西岸城市。
這是加拿大最受歡迎的甜點之一。原本的食譜是不需要烘烤的，質感就好像朱古力餅。
亞洲人可能會覺得比較甜膩，所以我做了一個蛋糕版本。用原本的材料，
加了一個蛋糕底，可以中和Nanaimo bar的甜度，同時又保留朱古力的濃香味道。

製作蛋糕

1 製作脆餅底
將溶化的牛油與消化餅乾碎混合，再倒入蛋糕紙杯的底部，用一小匙壓平至約4mm厚。

2 混合蛋糕糊和其他材料
準備一份基本朱古力蛋糕糊，拌入椰絲、核桃(杏仁)。

3 入爐烘焙
將麵糊慢慢倒入蛋糕紙杯，約2/3滿。用170度烘烤20-30分鐘或至蛋糕表面呈金黃色。

4 冷卻
完成後，轉移到散熱架上。等蛋糕完全冷卻後才裝飾。

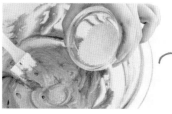 

裝飾步驟

填入吉士奶黃醬
在蛋糕中間挖一個小孔，填入吉士奶黃醬。

使用開縫星齒唧嘴
預備1個唧袋，放入開縫星齒唧嘴。然後拌入朱古力醬，往下推實，紮緊袋口，在蛋糕面上唧上花紋。

使用核桃、紙旗作裝飾
在蛋糕上撒一些核桃(杏仁)，再插上印有卡通圖案的紙旗作裝飾。

材料 Ingredients

脆餅底
40克無鹽牛油，在室溫下
80克消化餅乾(壓碎)

蛋糕
一份基本朱古力蛋糕
30克黑朱古力(熔化)
10克核桃/杏仁
35克椰絲

裝飾
開縫星齒唧嘴
1個唧袋
一份朱古力醬
吉士奶黃醬
椰絲
核桃/杏仁
卡通紙旗裝飾

For Cupcake

Bottom layer
40g unsalted butter, at room temperature
80g digestive biscuits (crumbs)

Cupcake Layer
1 portion of Basic Chocolate Cupcake Batter
30g dark chocolate chips melted and folded in batter
10g almonds/walnuts
35g desiccated coconut

For Decoration
Open Star Piping Tip
1 piping bag
1 portion of Chocolate Ganache
devon custard
desiccated coconut
walnuts/almonds
cartoon paper toppers

Cupcake Steps

1 For the bottom crust: Melt the unsalted butter and add in the digestive biscuits. Press firmly at the bottom of each cupcake liners about 4mm thick.

2 For the cupcake: Fold in the coconut and walnuts/almonds into 1 portion of Basic Chocolate Cupcake Batter.

3 Pour into cupcake cups, 2/3 full. Bake for 20-30 minutes or until golden brown.

4 Transfer to cooling rack. Cool down completely before frosting.

Decoration Steps

1 Cut a hole in each cupcake. Fill centre with devon custard.

2 Place the Open Star Piping Tip into a piping bag. Place the Chocolate Ganache into the piping bag and pipe onto each cupcake.

3 Sprinkle desiccated coconut and walnuts on top. Decorate with cartoon paper toppers.

薄荷朱古力蛋糕
Mint Chocolate Cupcakes

薄荷是好香口的！在加拿大的時候，我弟弟最愛吃一款薄荷朱古力，但是現在已經停產，市面上再買不到。我們常常懷念這朱古力的味道，所以我做了這款薄荷朱古力cupcake！這款cupcake由朱古力蛋糕底加上薄荷朱古力碎，再配上又香又滑的薄荷牛油糖霜組成。這蛋糕香濃可口，讓我們可以重拾已經失去很久的味道。

材料 Ingredients

蛋糕
一份基本朱古力蛋糕麵糊
1/2 茶匙薄荷香精油
30克薄荷烘焙朱古力粒
（可選）

薄荷牛油糖霜
一份基本的牛油糖霜
1/2 茶匙薄荷香精油
1-2 滴綠色食用色素

裝飾
圓口裱花唧嘴
1個唧袋
薄荷朱古力粒
黑朱古力(已溶)
綠色裝飾彩砂糖

For Cupcake
1 portion of Basic Chocolate
   Cupcake Batter
1/2 teaspoon peppermint
   essence
30g peppermint baking
   chocolate chips (optional)

For Peppermint Buttercream
1 portion of Basic Buttercream
1/2 teaspoon peppermint
   essence
1-2 drops green food coloring

For Decoration
Round Piping Tip
peppermint chocolate chips
dark chocolate (melted)
green color sanded sugar

製作蛋糕

1 混合薄荷香油和朱古力麵糊
將薄荷香油拌勻入一份預先預備好的基本朱古力蛋糕糊。

2 拌入朱古力粒
拌入可烘焙薄荷朱古力粒，拌勻。

3 入爐烘焙
將麵糊慢慢倒入蛋糕杯，約2/3滿。用170度烘烤20-30分鐘或至蛋糕表面呈金黃色。

4 冷卻
完成後，轉移到散熱架上。等蛋糕完全冷卻後才裝飾。

製作薄荷牛油糖霜

拌入薄荷香油
將薄荷香油拌勻入一份預先準備好的牛油糖霜裏。

加入綠色色素
加入1-2滴綠色食用色素，拌勻。

打發
打發至鬆軟滑身。

裝飾步驟

使用圓口裱花唧嘴
拿1個唧袋，放入圓口裱花唧嘴。將薄荷牛油糖霜放入唧袋，往下推實，紮緊袋口，在蛋糕表面唧上花紋。

用朱古力粒作裝飾
在糖霜上淋上黑朱古力醬，撒一些薄荷朱古力粒、綠色裝飾彩砂糖作裝飾。

Cupcake Steps

1 Add in the peppermint essence to 1 portion of Basic Chocolate Cupcake Batter.

2 Add in the peppermint chocolate chips.

3 Pour into cupcake cups, 2/3 full. Bake for 20-30 minutes or until golden brown.

4 Transfer to cooling rack. Cool down completely before frosting.

Peppermint Buttercream Steps

1 Add in the peppermint essence to 1 portion of Basic Buttercream.

2 Add in 1-2 drops of green food coloring.

3 Beat until well blended and smooth.

Decoration Steps

1 Place the Round Piping Tip into a piping bag. Place the Peppermint Buttercream into the piping bag and pipe onto each cupcake.

2 Sprinkle with peppermint chocolate chips, green color sanded sugar or mint chocolate on top.

蘋果金寶蛋糕
Apple Crumble Cupcakes

我在加拿大長大，對蘋果金寶情有獨鐘，每次見到餐牌上有
這個甜品，我都一定會點。我很喜歡暖暖的蘋果肉，
有很香的肉桂味，再加上烤得脆脆的口感，
實在是美味極了！最後千萬不能忘了冰淇淋這個「最佳拍
檔」！這款蘋果金寶cupcake就是升級版的蘋果金寶，因為又多
了一層軟綿綿的蛋糕口感。

外酥裏嫩，
好好味！

材料 Ingredients

蛋糕

一份基本雲呢拿牛油蛋糕麵糊

1茶匙肉桂粉

一個青蘋果（切粒）

金寶

50克麵粉

30克紅糖

35克無鹽牛油（室溫）

For the cupcake

1 portion of Basic Vanilla Cupcake Batter

1 teaspoon grounded cinnamon

1 green apple, chopped

For the crumble

50g plain flour

30g brown sugar

35g unsalted butter (room temperature)

製作蛋糕

1 混合肉桂粉和蛋糕麵糊

將肉桂粉加入一份預先備好的基本雲呢拿牛油蛋糕糊裏面，拌勻。

2 加入青蘋果粒

青蘋果切粒，拌入麵糊裏。

3 倒入蛋糕杯

慢慢倒入蛋糕杯，約2/3滿。

製作金寶

混合麵粉、紅糖、牛油

將麵粉和紅糖混合。再加入切成細方塊的牛油。

搓拌

用手將材料摩擦混拌，至呈麵包糠碎粒狀。

平鋪在蛋糕頂部

將碎粒鋪在每個蛋糕的頂部。

烘焙

用170度烘烤20-30分鐘至碎粒呈棕色，蛋糕表面呈金黃色。完成後，可配上雪糕。

Cupcake Steps

1 Add the grounded cinnamon to the Basic Vanilla Cupcake Batter.

2 Cut the apple into pieces and fold it into mixture.

3 Pour into cupcake cups, 2/3 full.

Crumble toppings Steps

1 Mix the flour and sugar. Cut the butter in cubes.

2 Rub in the butter using your fingers to mix into bread crumbs texture.

3 Sprinkle crumble on top of each cupcake.

4 Bake for 20-30 minutes until crumbs are brown and cupcake is golden brown. Served with ice cream (optional).

小杯子裏嘆世界 Wonderful World of Cupcakes

作者　Author
戚黛黛　蒙順意　Priscilla Chi　Monica Mong

編輯　Editor
黃雯怡　Annie Huang

攝影　Photographer
許錦輝　FaiDias Hu

化妝師　Makeup Artist
嚴芷瑩　Michelle Yim

封面設計　Cover Designer
吳明煒　MW NG

版面設計　Designer
何秋雲　Sonia Ho

出版者　Publisher
萬里機構・飲食天地出版社　Food Paradise Publishing Co., an imprint of Wan Li Book Co Ltd.
香港鰂魚涌英皇道1065號東達中心1305室　Room 1305, Eastern Centre, 1065 King's Road, Quarry Bay, Hong Kong.
電話　Tel: 2564 7511
傳真　Fax: 2565 5539
網址　Web Site: http://www.wanlibk.com

發行者　Distributor
香港聯合書刊物流有限公司　SUP Publishing Logistics (HK) Ltd.
香港新界大埔汀麗路36號中華商務印刷大廈3字樓　3/F, C & C Building, 36 Ting Lai Road, Tai Po, N.T., Hong Kong.
電話　Tel: 2150 2100
傳真　Fax: 2407 3062
電郵　E-mail: info@suplogistics.com.hk

承印者　Printer
凸版印刷(香港)有限公司　Toppan Printing Co (HK) Ltd.

出版日期　Publishing Date
二〇一五年六月第一次印刷　First Printing in June 2015

ISBN　978-962-14-5720-2

萬里機構　　　萬里 Facebook　　　myCOOKey.com

我的美食印象